THE TWENTY-FIRST-CENTURY MEDIA INDUSTRY

Studies in New Media

Series Editor: John Allen Hendricks,
Stephen F. Austin State University

This series aims to advance the theoretical and practical understanding of the emergence, adoption, and influence of new technologies. It provides a venue to explore how New Media technologies are changing the media landscape in the twenty-first century.

TITLES IN SERIES:

The Twenty-First-Century Media Industry: Economic and Managerial Implications in the Age of New Media, edited by John Allen Hendricks

THE TWENTY-FIRST-CENTURY MEDIA INDUSTRY

**Economic and Managerial Implications
in the Age of New Media**

Edited by
John Allen Hendricks

LEXINGTON BOOKS
A Division of
ROWMAN & LITTLEFIELD PUBLISHERS, INC.
Lanham • Boulder • New York • Toronto • Plymouth, UK

Published by Lexington Books
A division of Rowman & Littlefield Publishers, Inc.
A wholly owned subsidiary of The Rowman & Littlefield Publishing Group, Inc.
4501 Forbes Boulevard, Suite 200, Lanham, Maryland 20706
http://www.lexingtonbooks.com

Estover Road, Plymouth PL6 7PY, United Kingdom

British Library Cataloguing in Publication Information Available

Library of Congress Cataloging-in-Publication Data

The hardback edition of this book was previously cataloged by the Library of Congress as follows:

The twenty-first-century media industry : economic and managerial implications in the age of new media / edited by John Allen Hendricks.
   p. cm. — (Studies in new media)
 Includes bibliographical references and index.
 1. Mass media—Technological innovations. 2. Digital media—Economic aspects. 3. Digital media—Social aspects. I. Hendricks, John Allen.
 P96.T42T89 2010
 302.23—dc22

2010013283

ISBN: 978-0-7391-4003-1 (cloth : alk. paper)
ISBN: 978-0-7391-4004-8 (pbk. : alk. paper)
ISBN: 978-0-7391-4005-5 (electronic)

♾ ™ The paper used in this publication meets the minimum requirements of American National Standard for Information Sciences—Permanence of Paper for Printed Library Materials, ANSI/NISO Z39.48-1992.

Printed in the United States of America

As always, because of their continued love and support, my achievements are dedicated to Stacy, Abby, and Haydyn.

Contents

Part II: Implications of New Media Technologies

Figures and Tables

Figures

Tables

Foreword

THE PUBLICATION OF THIS EDITED VOLUME, entitled *The Twenty-First-Century Media Industry: Economic and Managerial Implications in the Age of New Media,* is an important step in the evolution of scholarship in the field of media management. The editor of this work, John Allen Hendricks of Stephen F. Austin State University, has assembled a strong group of scholars to offer some new perspectives and ideas on management in a changing media environment, involving both traditional and new media. Professor Hendricks and his collaborators are to be congratulated on their efforts.

Media management is a sub-field of the broader field of management, but the study of management is relatively young, at least compared to other disciplines. Management studies emerged slowly in the twentieth century, in large part due to a global shift from an agrarian to an industrial society in most developed nations. As the manufacturing of products became the primary activity in the Industrial Revolution, companies began to explore ways to increase productivity. In some cases, industrial psychologists were hired to help companies study and improve assembly plants and methods of production. These early efforts became known as the classical school of management.

In the 1930s and 1940s the study of management began to shift by focusing on the worker and his/her motivations and needs. By the 1960s, the modern era of management was introduced, exemplified by such milestones as the introduction of management by objectives, and Theory X and Y approaches. In the 1970s and 1980s, Total Quality Management, Theory Z, strategic management, and the study of leadership would make new contributions to the study of management, with some of these applied to the media industries.

It was during the behavioral era of management that research involving the media industries debuted. Most of the research on media management can be traced to the late 1940s with radio, followed by studies on newspapers and television in the 1950s. The introduction of cable television in the 1970s added a few books and studies on differences in managing a cable TV system. In the 1990s and the first decade of the twenty-first century, convergence became the buzzword and scholars began to examine the phenomenon mostly in newsrooms, where convergence was most likely to take place.

The problem is that during the latter part of the twentieth century, the media industries began to experience fundamental change, but our study of management remained stuck in a 1970s model. Most of the management research focused on single industries or single levels, and failed to acknowledge the impact of technology, globalization, economics, and social forces. Media industries were forced to adopt the Internet in the 1990s, and were slow to recognize the importance of blogs, social media, and tweets in the twenty-first century. Contemporary media managers are scrambling to try and figure out what to do with a range of possible digital platforms available to distribute content, how to generate new revenue streams from these platforms, increasing fragmentation and declines in audience viewing and listening, and the best ways to brand and market content.

The Twenty-First-Century Media Industry does not provide answers to all of the questions associated with this chaotic marketplace, but it does make an effort to explore some of the new management challenges that badly need to be addressed. Scholars in this volume look at a number of different industries including radio, television, newspapers, and music. Other chapters examine the impact of specific technologies such as the digital video recorder (DVR) and the mobile phone. Still other chapters address broader topics of change, adaptability, innovation, and distribution which can be applied to all segments of the media industries.

The Twenty-First-Century Media Industry is a welcome addition to the literature of the field as it attempts to deal with a fast-moving target—the evolving media marketplace. Students, practitioners and policymakers should find the content both compelling and thought-provoking, while presented in a format that is easily digested and understood. Let's hope this work will be an inspiration and catalyst for more new explorations on media management in the twenty-first century.

Alan B. Albarran
The University of North Texas
January 2010

Acknowledgments

I WISH TO THANK EACH AUTHOR WHO CONTRIBUTED an essay to be included in this book. Without their diligent work and enthusiasm for seeking a better understanding of the changing media landscape in the twenty-first century, this book would not have come to fruition. I am fortunate to have the support and understanding of an academic institution that values and promotes scholarly and intellectual pursuits such as this project. Thus, I would like to express appreciation to Stephen F. Austin State University in Nacogdoches, Texas. Appreciation is also extended to Deborah M. Smith, administrative assistant at Stephen F. Austin State University, for serving as "gatekeeper" on days that were needed to focus attention on research, writing, and editing. Also, appreciation is extended to Mark Rasko, senior secretary, who is a computer wizard and enthusiastically offered quick and knowledgeable assistance when the computer was not cooperative. Appreciation is also extended to my colleague Dr. Gary Mayer for his keen editing skills and to my graduate assistant, Kimberly Squyres, for her assistance with the project. Appreciation is also extended to Dr. Alan Albarran for his assistance with the book proposal and his participation in this project. In many ways, his work in media management inspired this project. I wish to thank Rebecca McCary, assistant editor at Lexington Books, for her support of this project and for being so diligent and conscientious with quick answers and help as issues arose. I want to thank Jehanne Schweitzer, senior production editor at Rowman & Littlefield Publishing Group, for her professionalism and assistance throughout the book's production process. Additionally, appreciation is extended to Michelle Kelly, Senior Vice President for Event Marketing, and Bart Stringham, Senior Vice

President and Corporate Counsel, of the National Association of Broadcasters for their assistance in acquiring permission to use the graphic for the book's cover. Lastly, and most important, I wish to express appreciation for the continued support and love of my wife, Dr. Stacy Nason Hendricks, and my children, Abby and Haydyn.

I

CHANGE: TECHNOLOGY, ECONOMIC IMPLICATIONS, AND CONSUMER BEHAVIORS

1

New Media

New Technology, New Ideas or New Headaches

Susan Smith and John Allen Hendricks

THE MEDIA INDUSTRY IS DYNAMIC AND INNOVATIONS are constant. Change is more pervasive and happening more quickly than at any other time in the history of the media industry. Technological advances across all forms of media are bringing about a revolution. Huge improvements in the gathering, production and delivery of content are occurring almost daily, and traditional media systems are attempting to keep up with the rapid changes and improvements that new media technologies bring.

Generally thought of as technological advancements, "new media" is influencing every aspect of the traditional media world. But is that influence always positive? New technology and traditional media have always had a somewhat uneasy alliance. Sometimes this alliance has created challenges.[1] The challenges come in many forms. For example, while content acquisition and distribution may be easier today, many media outlets are struggling with the financial burden that "going digital" requires. Some media organizations enthusiastically embrace new technology while others do not believe the financial burden is worth the risk. Some organizations are content to allow others to blaze the trail and sit on the sidelines to see if embracing the new technology is actually worth the risk. It is a tenuous relationship that has existed in the media throughout history.

Defining New Media

You might know the term "new media" by the name "emerging media." Or maybe you know it by the alias "digital technology." Or, better yet, maybe you call it computer-mediated communication. Whatever you call it, it surrounds you. New media bombards you. It is everywhere. Think about it. Think about your typical day. You awaken to an alarm on your mobile phone. The music from your MP3 player helps you welcome the day. After you roll out of bed, you check your e-mail on your computer, and then you flip to the morning news on your high definition television. Then it is time to download a podcast to listen to as you commute to work. There is road construction so you turn on your global positioning system to help navigate around it. You need to call your study group to let them know you are running late, so grab that cell phone and text them or call them. After a day spent working on the computer with digital technology, you head home. You have decided to relax tonight, so you are going to watch your favorite program, recorded the night before, on your digital video recorder (DVR). Or maybe you will check out a movie online. A full day surrounded by media, but is it truly "new" media?

If you define new media at face value, then it is an idea that has been around as long as there have been spoken and written languages. New media, media that is new, does that mean that when Moses brought the word of God down from Mount Sinai on the tablets that he was doing it with new media? Or when Johannes Gutenberg invented the printing press and began using it as a tool to spread information, was he doing so on new media? Both examples were technological improvements to the way that content was delivered previously. But to call them new media goes too far. There is an old joke that by the time we all figure out what new media really is it will be old. So before it gets old, let's look at some of the ways that it has been described.

Communication scholar Marshall McLuhan first used the term "new media" in 1953. McLuhan was futuristic in his use of the term to talk about the technology of communications such as electronic information gathering and global reach.[2] But the term really emerged in the late 1990s when it began to be used as an all-encompassing description for emerging and digital technologies.[3] Because of that, it is a concept that has been associated with the spread of information digitally. Until the 1980s, media relied on delivery systems like print and broadcast that were primarily analog in nature. Use of the term "new media" helps to differentiate the digital media from the old media, the analog signals. Lev Manovich suggests that in the same way the printing press and photography impacted the development of modern society and culture during the centuries they were introduced, new media is revolutionizing culture today, in the shift to "computer-mediated forms of production, distribution and communication."[4]

The latter half of the twentieth century saw an explosion in the communication industry. Personal computers, satellites, cable television, cell phones, digital and high definition television, DVDs and the World Wide Web are just a few of the technologies that have become closely associated with new media. It has been a true revolution in the communication industry. Even old media companies have begun to embrace new media, whether out of a desire to embrace the new technologies, or out of necessity, few old media companies exist in their previous form. But the term new media has also enveloped even more than just digital communication. It has become associated with converged, computerized, networked, interactive and compressible technologies and information. If we go with this all-encompassing definition, we are primarily talking about only the technology associated with new media. For many, the definition goes even further. It is not just the technology, but it is the way in which we interact with the technology that truly makes it new media.[5] For others, it is the convergence of the new technologies, the way in which content is used across multiple media platforms; it has come to be known as the "technological, industrial, cultural, and social changes" to, and within, media.[6] In a more basic description, new media refers to a wide range of changes in media production, distribution and use.[7] Still others identify it by the new technologies and the interdependence they offer the user. The short answer is that it is all these things, but to truly understand new media, we need to look at each of these ideas.[8]

The Technology of New Media

Technology has become the word most closely associated with new media. Those things that we as consumers consider new media are often the latest "must have" pieces of equipment that make life easier and more entertaining. Imagine your life without satellite television, digital video recorders, iPods, cell phones or even computers. You cannot. And before those "must have" pieces of equipment, it was something else that we could not live without: a videocassette recorder (VCR), a television or a radio. But are these items really new media? The concept of new media suggests that it replaces an old medium; however, it does not work that way.[9] When television came into existence, it did not replace the medium of radio. When video began being used on the Internet, it did not replace the medium of television. "In the 1990s, rhetoric about the coming digital revolution contained an implicit and often explicit assumption that new media was going to push aside old media, that the Internet was going to displace broadcasting, and that all of this would enable consumers to more easily access media content that was personally meaningful to them."[10] It is not the media that is changing; it is the delivery systems that change. The tools that we use to access media change as technology changes.[11]

Other new media scholars believe that it is more than just the delivery system; they believe that the computer is the one piece of technology that revolutionized the media industry. Most every technology that is closely associated with new media is just as closely associated with computer technology. But accepting the idea that computer-mediated technology is new media is limiting.[12] Lev Manovich says that photographs by themselves are not new media, but photographs displayed on a DVD are new media. Text stories distributed on a website are new media, while the same text in a newspaper is not. Why? According to Manovich it has to do with the computerization of the media either on the DVD or on the website. What makes the computer-mediated product new media is the ability to break it down into computer data. The ability to break down graphics, moving images, sounds, and text into numerical data through computerization makes it all new media.[13]

The advent of the personal computer in the mid-1980s began a flood of ideas that challenged the existing assumptions about media, culture and technology and opened the floodgates of constant and rapid technological innovation that has come to be associated with "new" media.[14] John Pavlik agrees that the digital transformation of media is what has become accepted as "new media."[15] But Lev Manovich and John Pavlik both, like many others, believe that looking simply at the storage and distribution aspects of computerization as the thing that puts the "new" in new media is not a complete view. Computerization also impacts the acquisition and production of content as well as the storage and distribution. Pavlik identifies twelve dimensions of the transformation of media in the digital age. The convergence of media looks at not only delivery and storage but also the technology of production, the acquisition, the audience, the producers, the content, the distribution, the law and ethics of new media, the innovations, the next generation of media consumers and the business and financing of new media.[16]

Others do not break it down quite so much but instead acknowledge that "new" media can refer to a wide range of changes in media production, distribution and use, which are more than technological; they are also textual, conventional and cultural.[17] So what does that say about new media? It tells us that it is not just the technology of the medium, but rather the way consumers and audience members interact with the medium. It is what they see and how they see it as well as the manner in which it is financially supported. So if we return to the introduction of Gutenberg's printing press in the fourteenth century we can now see why it would not be considered "new media." While it had a great technological impact on the process of communication at the time, and despite the fact that it had a great impact on cultural communication, it affected only one stage of cultural communication, the distribution of media.

In today's "new media" revolution, the technological development of computer-mediated communication impacts all stages of communication including, as was indicated before, acquisition, manipulation, storage and distribution. It also affects all aspects of media.[18] Without the technology, "new media" as we know it would not exist. However, as Lev Manovich and John Pavlik and many others suggest, it is not enough for media to exist as pieces of digitized, numeric information to allow it to be called new, it is also about how the digitized, numeric pieces of information are interacted with in today's society.

New Media in a Global Society

New media is about more than just the transfer of information across digital highways; it is about the way in which those informational digital highways are traveled, the routes that are taken and the way in which you interact with the highway to get you where you are going. Imagine driving a highway without the use of a Global Positioning System (GPS). It is not hard to imagine. Not too many years ago, if you were lost on a highway in an unfamiliar area, you either pulled out an "old" medium, a map, or you stopped and asked for directions. Today's technology provides us with the ability to use GPS to pinpoint our location and see where we are in reference to where we want to go. But it is not enough to just have the GPS. What truly makes it a new medium is our ability to interact with it and work through our issues of being lost and finding our way to our ultimate destination. Without something that connects these roadways, you might as well just stay at home, because navigating the digital highway will be almost impossible.

The advent of communication satellites opened a whole new highway of accessibility to mediated information. Until then, information traveled primarily within small geographic areas. When communication satellites began carrying pieces of information across the country and across the world, media suddenly had the ability to impact cultures globally and conversely be impacted by global cultures.[19]

Just as communications satellites change the face of media, the Internet has impacted the global reach of media. Information and entertainment media from the United States is reaching parts of the world that in the past would be cost prohibitive. In the same way that media from the United States is reaching other parts of the world, media from other countries and cultures is now reaching the United States. This global infiltration of media is not just about the spreading of a particular message; it is also about the way in which other parts of the world are using media. The differences in media relationships and media usage are impacting the development of these "new media" as well as the delivery, the consumption, the acquisition, the regulation, and virtually all aspects of new media. According to Henry Jenkins, the consumers' active participation with and within the media is necessary for convergence to exist. Jenkins asserts: "This circulation of media content—across different media systems, competing media economies, and national borders—depends heavily on consumers' active participation."[20] Further, Jenkins believes that convergence is more than just a technological process, but rather a way to encourage people to seek out new information and make connections with all kinds of widely dispersed media content.[21]

Nowhere is the globalization of media more apparent than in the news industry. Without "new media," news traveled slowly and often only within an audience member's own limited geographic region. Communication satellites changed the accessibility of news. Information began reaching audiences quickly and from parts of the world that had at one time been impossible to reach. Both the newspaper industry and the television industry benefited from the globalization of news and information.[22]

The globalization of news and information continued to grow in the late twentieth century when the Internet changed the face of delivery and accessibility of news and information to the audience. Not only could users get information from parts of the world that at one time was impossible to obtain, but they were now getting it at a seemingly instantaneous speed. But even the news media realized that it is more than the delivery system that plays a role in making "new media" new. John Pavlik suggests that new technology provides opportunities for the news industry in four areas:

1) More efficient ways for journalists and other media professionals to do their jobs.
2) Transforming the nature of storytelling and media content in general in potentially positive and engaging ways, especially with younger audiences. (e.g., podcasts, mobile delivery, and greater variety of distribution of information).
3) Implications for management, structure and culture of media organizations.
4) Transformation of the relationships between news organizations and their many publics (e.g., audience, sources, funders, regulators and competitors).[23]

While the technology of media plays an important role in making "new media" new, even in the news industry, it is the interaction with and within the media that helps to truly define it as "new."

The Economics of New Media

If we work on the premise that traditional media do not disappear but are simply improved or updated by "new media," then what does "new media" have in store for newspapers? Arguably one of the most visible examples of traditional media that has been adversely impacted in many ways by "new media" is the newspaper industry. Newspaper circulation losses continue and according to the annual report on American journalism by the Pew Project for Excellence in Journalism, those circulation losses are accelerating.[24] Advertising is dropping in the printed media and earnings are plummeting. It has many of these old media companies looking for new economic models involving "new media."

Business models supported by online advertising remain questionable. Online versions of newspapers have supported advertising in the past, but never enough to cover the costs of newsgathering.[25] Certainly there are news organizations that have eliminated the print versions of their newspapers and are now delivering the news in an online-only format. But can these organizations survive? They certainly hope so. These online editions are trying to figure out ways to monetize their content, which has generally been offered to the public for free. Some believe that an advertising model can work at a break-even or slightly profitable level. No matter which model is used, there is still a gap between the funding and revenue of these news organizations versus the expenses and what will inevitably suffer is the news that the consumer receives.[26]

Not only newspapers but also traditional broadcast entities are looking toward new media to generate additional revenue. The Internet and more recently mobile and second-screen delivery methods are a focus of many newsrooms and entertainment divisions of broadcast companies. While users show no interest in paying for content on the Internet, they are willing to pay for mobile content. Apps, or applications, for the iPhone are huge money-makers, and Amazon's Kindle, offering newspaper subscriptions, appears to have revenue potential for news organizations.[27] Add video on demand, blogs, citizen journalism and podcasts among the digital "new media" tools that old media companies are utilizing to cultivate a new, more technologically savvy audience and figure out how to get the audience to pay for them and these news organizations may just be in business!

Traditional Media versus New Media

There exists a debate among media industry observers as to whether new media technologies will drastically alter the traditional media landscape and perhaps serve to bring about its demise. In 2007, a report was released that warned "the conflict between traditional media and new media is seeing the emergence of a media divide that could erase hundreds of billions of dollars in revenue from the bottom line of the world's leading media companies."[28] Further, the study found "new forms of media will grow from 11 percent to 18 percent of total market revenues over the next four years, with new media growing at five times the rate of traditional media."[29] In contrast, Robert S. Boynton states: ". . . a number of recent developments suggest that new media may actually be the salvation of old media; that online newspapers, Webzines, and e-books could preserve and extend the best aspects of the print culture while augmenting it with their various technological advantages. If this is true, then the future of old media is in embracing the new . . ."[30] Boynton suggests that traditional media has an advantage that new media does not have: "Old media has an advantage it must nourish: credibility."[31]

Discussion among media industry insiders also surrounds content and the specific type of content twenty-first-century consumers find appealing. Steven Canepa, IBM vice president of global media and entertainment industry, asserts: ". . . it's not just providing compelling content but providing compelling experience and part of the experience may be something that extends beyond traditional content."[32] The future of traditional media is impacted by the technology that provides consumers with the content they want when they want it. The consumer appeal of not being tied down to specific times to watch a television show or listen to a certain radio show is evident in the adoption of new media technology. The adoption and acceptance of the Internet by the general public is growing five times as fast as the acceptance rate of television when it first emerged as new technology.[33]

New Media Technology and Emerging Business Models

Online video content continues to grow and mergers continue to occur in boardrooms. In 2006, Sony Pictures Home Entertainment and Time Warner's Warner Brothers Entertainment partnered with the online video sharing site GUBA while NBC partnered with YouTube to provide a wide range of video content online.[34] It was observed: "Add up the venture capital dollars funding online video startups, the technology advances, the willingness of established players like ABC, CBS, and NBC to try new distribution models and the increasing web viewership, and it's clear that the video market is at an inflection point . . ."[35] Despite impressive attempts, viable business models have yet to emerge. Online video providers have yet to find embedding advertising in video to be profitable. Part of the problem with this business model is that online video content is not like traditional media content. Online video content is primarily generated by users and uploaded to sites like YouTube and Guba. However, the Sony, Time Warner, and NBC deals are likely to form some type of viable business models. Moreover, Kendall Whitehouse, a Wharton School professor, asserts that unlike in the past, ". . . advertising may not be a holy grail. Consumers may revolt when ads start appearing in videos . . ."[36]

The Warner Brothers/Guba deal permits Warner Brothers programs to be sold from the Guba website, which may prove to be a profitable business model as the details are worked out. In contrast, other media observers assert that television will remain the dominant medium that will deliver content to consumers. Forrester Research predicts ". . . a decade-long evolution that will ultimately result in most programming delivered on-demand with targeted ad messages based on location and behavior . . ."[37] Since 1984, television network news viewership has dropped 37.8%.[38]

In addition to dwindling audiences, advertising revenue is also dropping. Particularly, revenue is expected to continue dropping at local television stations. As a result, television executives are moving some shows to cable networks and away from traditional terrestrial television stations in order to generate steady revenue from cable subscribers.[39] Stations are looking for new revenue streams and are sending their signals and content to cell phones and other mobile devices and selling more online advertising.[40] Another observer notes television programming will eventually migrate to the Internet and will be successful if the "pay-per-view" model is used. Consumers prefer programming when it is convenient for their schedules. Jim Edwards, of Practical eCommerce, notes: "I hate to make this overly simplistic, but bottom line: an effective online 'TV Station' only needs a basic website and the ability to allow 'viewers' to download or stream video files."[41] In early 2009, Santeon Corporation launched a computer software platform system that offers "viewers a much more interactive and personalized experience through content search, pay-per-view and pay-to-view browsing or surfing as well as other online interactive features with TV 2.0 collaboration. The platform also has the ability to create linear and non-linear scheduled programming with highly customizable template-based rich user interfaces."[42] Additionally, Yahoo has partnered with Samsung, Toshiba, and LG to create a Yahoo Widget Engine that displays a menu bar across the bottom of the television screen that will enable the user to access the Internet while watching television. Yahoo has labeled this as "cinematic television."[43]

Even the Public Broadcasting Service (PBS) has created an online portal that offers consumers all of its best programs. PBS posts thousands of hours online. Jake Coyle states: "PBS already has a channel on Hulu, a joint venture of News Corp. and General Electric's NBC Universal, and it posts shorter clips on YouTube. But the new player will eventually include material from its 357 affiliates, offering an online destination for PBS viewers."[44] PBS hopes this move will generate a younger viewership for its content. Furthermore, Coyle observes: "In the network's ongoing efforts to shed an old stodgy image, PBS—which turns 40 this year—is also present in social networking sites like Facebook and Twitter."[45]

Technological advancements are also reaching and affecting the radio industry at an impressive pace. One observer noted: "The digital revolution took its time getting to radio. Now it's exploding—and the big bang goes far beyond podcasting. As radio shows are turned into digital bits, they're being delivered many different ways, from Web to satellite to cell phones. Listeners no longer have to tune in at a certain time, and within range of a signal, to catch a show or a game."[46] There are a host of alternatives to traditional terrestrial radio including satellite radio such as SiriusXM, Yahoo Internet radio, and MSN radio. And, one observer notes: "It's possible to imagine people paying monthly fees to hear programming-on-demand on the phone, PC, or in the car. Listeners could buy a song they hear on the radio with the click of a button. Companies could sell subscriptions and place ads inside customized traffic information, weather reports, or sports tickers."[47]

Furthermore, podcasting is not only eating into the radio industry's listener base but also diminishing advertising dollars among traditional media outlets. Major companies like Volvo are sponsoring podcasts and traditional broadcasters are starting their own podcasts to allow consumers to choose the content they prepare and allow consumers to listen when it is convenient for their schedules. Similarly, the radio industry is moving toward new media technology slowly because of a lack of a clear business model. Nicole Garrison-Sprenger noted: "These companies, still very few in number, understand a large audience is moving to podcasts and that they should be there, too. But executives are treading lightly, since few podcasters have figured out how to profit from it and navigate the pitfalls."[48] There are several online sites that consumers can visit to find podcasts that appeal to their interests such as www.podcastalley.com, www.podcastpickle.com, www.podcastready.com, and www.epnweb.org.[49]

National Public Radio (NPR), as a result of consumer demand, began offering podcasts several years ago. NPR's podcasts regularly hold some of the top ranks on iTunes' most popular podcasts. As a business model, NPR has entered into agreements with its affiliates to sell and share the underwriting income.[50] NPR's decision to move toward podcasting was due to an effort to find a new business model. Mark Glaser states: "Previously, NPR's income was split evenly from fees paid for content by member stations (who raised money from pledge drives), and corporate and foundation underwriting spots. Podcasting gave NPR a new model for selling underwriting, and sharing the proceeds with stations."[51]

As NPR explored producing podcasts for distribution, it found that the most successful podcasts contained programming that did not previously have an outlet for consumers. NPR found that consumers did not want repurposed content. As NPR worked to improve its business model, it noted some obvious weaknesses. Mark Glaser observed: "As for making money or getting underwriting for podcasts, everyone agrees that there are a slew of issues to iron out—though there's a lot of potential. First off, anyone who sells advertising usually has to have metrics on the audience: who is listening, how often they listen, what's the demographic of listeners. These remain a mystery for podcasts, because there is no current way to track who actually listens to podcasts."[52] Creating a successful business model from podcasting is further complicated by the fact that consumers of new media technologies are not keen on the idea of intrusive advertising. Glaser notes: "NPR has an advantage because it is already well versed in using less intrusive ads in its radio and Web programming."[53]

The newspaper industry is also experiencing changes as a result of new media technologies. Since 1984, newspaper circulation has dropped more than 15%.[54] After more than 145 years of serving the Seattle area, the *Seattle Post-Intelligencer* is shifting all of its content to an online format. This followed Denver's *Rocky Mountain News*, after 150 years of service, and Arizona's *Tucson Citizen* closing after many years of service due to economic downturns. Donna Gordon Blankinship notes: "The newspaper industry has seen ad revenue fall in recent years as advertisers migrate to the Internet . . ."[55] Other prominent newspapers sought bankruptcy, including the *Los Angeles Times*, *Chicago Tribune*, and the *Philadelphia Inquirer*. Steven R. Swartz, Hearst Newspapers president, said the *Seattle Post-Intelligencer*'s online presence is ". . . an effort to craft a new type of digital business . . ."[56] At a gathering of technologically savvy observers of the changing media landscape, Steven Johnson, who is a science writer, compared newspapers in the age of new media technology "to old growth forests, saying that under the canopy of that aged ecosystem blogging, citizen journalism, Twittering and other Internet-age information sharing is taking root."[57] Further, Johnson "sees the future of news weaving together talents of professional journalists, bloggers, and people using social networking tools such as Facebook and Twitter to instantly tell what is happening around them."[58]

Generational Shift and Media Consumption

Interestingly, new media technologies are attracting younger consumers which does not bode well for traditional media outlets. The 18-24-year-old demographic is considered the Millennial Generation, those born after 1984, which is the largest generation in history.[59] The Millennial Generation is a technologically savvy generation that relies heavily on new media technologies to obtain information ranging from the news and weather to communicating with peers via text messaging and social networking. Accordingly, one study indicated that the average American spends more than five hours a day watching television, but 18-24-year-olds spend only 3.5 hours a day watching television.[60] Further results from the study indicated that 18-24-year-olds spend an average of 26 minutes per day with video game consoles compared to 6.5 minutes a day for adults over the age of 24. The Millennial Generation averages 43 minutes a day on mobile devices compared to an average of 20 minutes a day with consumers over the age of 24.[61] In a separate study, it was discovered that "22 percent of adults 18-24 and 14 percent of adults 25-34 use the Internet to get the news."[62]

New Revenue Outlets

Media managers are forced to look to the Internet to distribute content and generate new revenue due to the impressive numbers of consumers that are migrating from traditional media outlets to new media outlets online. In 2007, one study indicated that, "the global entertainment & media sector is forecast to be worth $2 trillion in five years time, growing at a compound annual growth rate of 6.4%. Around 50% of this growth is expected to originate online, with forecasts of 540m broadband Internet subscribers driving migration to digital formats and the rise of user generated social media having an adverse impact on competing revenue streams and content fragmentation."[63] Also, observers note that online content is blurring the definition of media. Abbey Klaassen asks: "Does a media site create content with the goal of selling ads or subscriptions to pay for it? Or can we now define as a medium any site that aggregates an audience through other means—e-commerce or lead generation—then it turns around and sells to advertisers? In short, if Papa John's promotes Blockbuster and Coca-Cola on its site, does that make the pizza purveyor a media company?"[64]

Even public relations firms are struggling with how to adapt to new media technologies. Traditional media served as "gatekeepers" and public relations firms worked with the "gatekeepers" to get their messages delivered to the viewers, listeners, and readers in a professional manner. New media technology has now placed the consumer in the powerful and influential position of generating information and content and leaping past the "gatekeeping" role altogether. A. C. Croft observed: "Many of the 'citizen journalists' manning various 'new' media can hardly claim the same experience, objectivity and credibility as a traditional print or electronic journalist. So their output often tends to lack objectivity and 'third party credibility.'"[65] New media technologies have changed several important aspects of the public relations industry: 1) Static press releases are antiquated now that blogs are mainstream and consumers have the ability to interact and comment on information, 2) To be successful at public relations, you must do more than just learn the new media technologies, you must become proficient with the new media technologies, and 3) Search engines need to be utilized by public relations firms to make sure that certain terms appear when consumer searches are conducted.[66]

Twitter

Technological advancements continue to play a role in the way in which humans communicate. One of the most recent technological advancements to capture the attention of consumers is Twitter. Consumers are able to establish a Twitter account and can post comments that contain 140 characters or fewer. Friends can subscribe to an individual's Twitter account as a "Follower" and read the 140-character comments known as "tweets." Ben Grossman, Editor-in-Chief, of the trade publication *Broadcasting & Cable*, asserts: "I don't see Twitter as anything more than a short-term fad nearing the end of its 15 minutes."[67] The consumer retention rate for Twitter is not as impressive as it is for social networking sites like Facebook. In the spring of 2009, Nielsen rating company found: "For every 10 people who join the service, six will stop using it within a few weeks. Nielsen was cautious in its assessment, but those numbers do not bode well for Twitter. Facebook and MySpace, by comparison, had retention rates upward of 70% at the same stages in their development."[68]

Future Implications and the Changing Media Landscape

This book explores the implications of the proliferation of new technologies and new media for the traditional media industry. The rapid pace at which new technology is being developed by manufacturers and adopted by consumers is unprecedented. This is creating a situation in which media managers are required to allocate both financial and personnel resources to stay abreast of new media technology and determine where it fits into the organization's current operation. To be successful in the twenty-first century, media managers must adapt quickly. Scott Davis asserts: "The marketing and consumer landscapes have been undergoing tremendous changes in the past five years. The trick to avoid getting left behind in this brave new world is to embrace the impact, implications and influence you and your organization may have by moving just a tad into the unknown."[69] Media managers are finding that making these decisions is not easy. One study found: "While 53% of the 100-plus marketers surveyed acknowledged new media's future importance to their marketing mix, 31% of them admitted they just don't know how best to use these new tools to meet business goals."[70]

For traditional media outlets, the implications of new media are far reaching. Media managers are creating new departments, or units, to embrace and implement new media technologies. New media technologies are being integrated into existing industry operations through media convergence, but in some situations new media are actually replacing traditional media operations both within organizations and by consumers. For example, most television news operations have websites that complement their traditional television news programs. Many consumers now choose to go directly to the television news website to read and watch the video of only the stories they are most interested in that day. Television stations have supplemented their revenue by selling advertising on their Web pages. Television stations are even utilizing podcasting technology to reach consumers.

In turn, media managers are finding themselves writing job descriptions and seeking employees with cutting-edge skills that were never required by traditional media outlets. This is only one of many examples of new media technologies having a tremendous effect on the media industry from both a financial and managerial perspective. Not all news is bad concerning new media and traditional media. One study found "that most news consumers prefer to use new media as a complement to print and television rather than as a substitute."[71] Not all industry experts believe that new media technologies will be the demise of the traditional media landscape. Douglas Ahlers and John Hessen predict: "While industry-sponsored research tends to view the audience as monolithic consumers of one media at the expense of others, the reality of the dynamic 'multichannel' media user is very much the norm."[72] Moreover, Rupert Steiner argues that traditional media and new media will coexist. He states: "it will be less about one versus the other and more about forging a symbiotic relationship."[73]

Notes

1. John V. Pavlik, *Media in the Digital Age* (New York: Columbia University Press, 2008), 1–4.

2. Benjamin Peters, "And Lead Us Not into Thinking the New Is New: A Bibliographic Case for New Media History." *New Media & Society,* (2009), http://www.columbia.edu/~bjp2108/blog/Peters%20NMS%202009.pdf (28 November 2009).

3. Peters, "New Media History." *New Media & Society,* (2009).

4. Lev Manovich, *The Language of New Media* (Cambridge, MA: MIT Press, 2001), 19.

5. Manovich, *The Language of New Media,* 39–51.

6. Henry Jenkins, *Convergence Culture, Where Old and New Media Collide* (New York: New York University Press, 2006), 3.

7. Martin Lister, Jon Dovey, Seth Giddings, Iain Grant and Kieran Kelly, *New Media: A Critical Introduction* (London: Routledge, 2003), http://books.google.com/books?hl=en&lr=&id=UkqomXHmoAEC&oi=fnd&pg=PP13&dq=%22Lister%22+%22New+media:+A+critical+introduction%22+&ots=_j-4BWOj0_&sig=R-0FnkB_EvVuRwU3n76RF1y06z8#v=onepage&q=&f=false (28 November 2009).

8. Dan Harries, preface to *The New Media Book* (London: British Film Institute, 2002), ix–x.

9. Jenkins, *Convergence Culture,* 13–14.

10. Jenkins, *Convergence Culture,* 5.

11. Jenkins, *Convergence Culture,* 14–15.

12. Manovich, *The Language of New Media,* 19.

13. Manovich, *The Language of New Media,* 20.

14. Lister, Dovey, Giddings, Grant and Kelly, *New Media: A Critical Introduction* http://books.google.com/books?hl=en&lr=&id=UkqomXHmoAEC&oi=fnd&pg=PP13&dq=%22Lister%22+%22New+media:+A+critical+introduction%22+&ots=_j-4BWOj0_&sig=R-0FnkB_EvVuRwU3n76RF1y06z8#v=onepage&q=&f=false (28 November 2009).

15. Pavlik, *Media in The Digital Age,* 8–11.

16. Pavlik, *Media in The Digital Age,* 9.

17. Lister, Dovey, Giddings, Grant and Kelly, *New Media: A Critical Introduction* http://books.google.com/books?hl=en&lr=&id=UkqomXHmoAEC&oi=fnd&pg=PP13&dq=%22Lister%22+%22New+media:+A+critical+introduction%22+&ots=_j-4BWOj0_&sig=R-0FnkB_EvVuRwU3n76RF1y06z8#v=onepage&q=&f=false (28 November 2009).

18. Manovich, *The Language of New Media,* 49.

19. Michele Hilmes, "Cable, Satellite and Digital Technologies," in *The New Media Book,* ed. Dan Harries (London: British Film Institute, 2002), 7–14.

20. Jenkins, *Convergence Culture,* 3.

21. Jenkins, *Convergence Culture,* 2.

22. Pavlik, *Media in The Digital Age,* 16–22.

23. Pavlik, *Media in The Digital Age,* 4–6.

24. The State of the News Media—An Annual Report on American Journalism, *PEW Project for Excellence in Journalism*, 2009, http://www.stateofthemedia.org/2009/narrative_newspapers_intro.php?media=4 (15 October 2009).

25. Glenn Fleishman, "Dual Perspectives, Future of Newspapers: Newspapers Founder, But Civic Journalism May Survive," *Wired*, 14 July 2009, http://www.wired.com/dualperspectives/article/news/2009/07/dp_newspaper_ars0714 (28 November 2009).

26. Fleishman, "Dual Perspectives, Future of Newspapers," http://www.wired.com/dualperspectives/article/news/2009/07/dp_newspaper_ars0714 (28 November 2009).

27. Douglas Wolk, "Dual Perspectives, Future of Newspapers: Profitless? Go Wireless," *Wired*, 14 July 2009, http://www.wired.com/dualperspectives/article/news/2009/07/dp_newspaper_wired0714 (28 November 2009).

28. Simon Canning, "Old vs. New May Cost Billions," *The Australian*, 15 February 2007, http://web.ebscohost.com/ehost/delivery?vid=8&hid=112&sid=1105a8e1-a 307-4 . . . (8 February 2008).

29. Canning, "Old vs. New May Cost Billions," *The Australian*, 15 February 2007.

30. Robert S. Boynton, "New Media May be Old Media's Savior." *Columbia Journalism Review*, July/August 2000, 29.

31. Boynton, "New Media May be Old Media's Savior." *Columbia Journalism Review*, July/August 2000, 34.

32. Simon Canning, "Old vs. New May Cost Billions," *The Australian*, 15 February 2007, http://web.ebscohost.com/ehost/delivery?vid=8&hid=112&sid=1105a8e1-a307-4 . . . (8 February 2008).

33. Carole M. Howard, "Technology and Tabloids: How the New Media World is Changing Our Jobs." *Public Relations Quarterly*, 45, no. 1 (Spring 2000): 8-12.

34. "Online Video: The Market is Hot, but Business Models are Fuzzy," Knowledge@Wharton, July 12, 2006, http://knowledge.wharton.upenn.edu/article.cfm?articleid=1519 (24 March 2009).

35. "Online Video: The Market is Hot, but Business Models are Fuzzy." *Knowledge@Wharton*, July 12, 2006, http://knowledge.wharton.upenn.edu/article.cfm?articleid=1519 (24 March 2009).

36. "Online Video: The Market is Hot, but Business Models are Fuzzy," *Knowledge@Wharton*, July 12, 2006, http://knowledge.wharton.upenn.edu/article.cfm?articleid=1519 (24 March 2009).

37. Brian Morrissey, "TV's Future Looks Like Web's Present: Forrester Says Targeted Ads and a Portal-Like Menu of Options are Coming to Your Set." *Adweek.com*, 27 August 2008. (24 March 2009).

38. Douglas Ahlers and John Hessen, "Traditional Media in the Digital Age: Data About News Habits and Advertiser Spending Lead to a Reassessment of Media's Prospects and Possibilities, *Nieman Reports*, (Fall 2005): 65–68.

39. "Local TV Stations' Uncertain Future." *MediaPost.com*, *The Wall Street Journal*, 11 February 2009, http://www.mediapost.com/publications/?fa=Articles.showArticle&art_aid=100112 (24 March 2009).

40. Sam Schechner and Rebecca Dana, "Local TV Stations Face a Fuzzy Future." *The Wall Street Journal*, 12 February 2009, http://online.wsj.com/article/SB123422910357065971.html (24 March 2009).

41. Jim Edwards, "Create Your Own Online TV Station," *Practical eCommerce.com*, 28 August 2006. http://www.practicalecommerce.com/articles/271-Create-Your-Own-Online-TV-Station (24 March 2009).

42. Gihan Morris, "Santeon Launches New Media Management Platform to Support Web/TV 2.0," *Reuters.com*, 21 January 2009. http://www.reuters.com/article/pressrelease/idUS113985+21-jan-2009+BW20090121 (23 March 2009).

43. Iain Thomason. "CES: Yahoo Bets on Internet TV as the Future," *vnunet.com*, 08 January 2009. http://www.vnunet.com/vnunet/news/2233520/ces-2009-yahoo-bets-internet-tv (24 March 2009).

44. Jake Coyle, "On the Net: PBS Releases its Programming Online," *Sherman, Texas, Herald Democrat*, 24 April 2009, D1.

45. Coyle, "On the Net: PBS Releases its Programming Online," *Sherman, Texas, Herald Democrat*, 24 April 2009, D1.

46. Heather Green, Tom Lowry, and Catherine Yang, "The New Radio Revolution," *BusinessWeek.com*, 3 March 2005, http://www.businessweek.com/technology/content/mar2005/tc2005033_0336_tc024.htm (24 March 2009).

47. Green, Lowry, and Yang, "The New Radio Revolution," *BusinessWeek.com*, 3 March 2005.

48. Nicole Garrison-Sprenger. "Business (Pod)casts for Clients." *Minneapolis/St. Paul Business Journal*, 5 August 2005, http://twincities.bizjournals.com/twincities/stories/2005/08/08/story2.html?t=printable (24 March 2009).

49. Traci Gardner, "Podcasts: The 21st Century Version of the Radio Show," *NCTE Inbox*, 22 July 2008, http://ncteinbox.blogspot.com/2008/07/podcasts-21st-century-version-of-radio.html (24 March 2009).

50. Mark Glaser, "Will NPR's Podcasts Birth a New Business Model for Public Radio?" *Online Journalism Review*, 29 November 2005, http://www.ojr.org/ojr/stories/051129glaser/print.htm (24 March 2009).

51. Glaser, "Will NPR's Podcasts Birth a New Business Model for Public Radio?" *Online Journalism Review*, 29 November 2005.

52. Glaser, "Will NPR's Podcasts Birth a New Business Model for Public Radio?" *Online Journalism Review*, 29 November 2005.

53. Glaser, "Will NPR's Podcasts Birth a New Business Model for Public Radio?" *Online Journalism Review*, 29 November 2005, http://www.ojr.org/ojr/stories/051129glaser/print.htm (24 March 2009).

54. Douglas Ahlers and John Hessen, "Traditional Media in the Digital Age: Data About News Habits and Advertiser Spending Lead to a Reassessment of Media's Prospects and Possibilities, *Nieman Reports*, (Fall 2005): 65–68.

55. Donna Gordon Blankinship, "Seattle Post-Intelligencer Newspaper Goes Web-Only," *Yahoo! Tech News*, 17 March 2009, http://tech.yahoo.com/news/ap/20090317/ap_on_hi_te/seattle_p_i (24 March 2009).

56. Blankinship, "Seattle Post-Intelligencer Newspaper Goes Web-Only," *Yahoo! Tech News*, 17 March 2009, http://tech.yahoo.com/news/ap/20090317/ap_on_hi_te/seattle_p_i (24 March 2009).

57. "Journalism Evolving, Not Dying: Science Author," *Breitbart.com*, 14 March 2009, http://www.breitbart.com/print.php?id=CNG.99c75c577fa1a698c1689adb0e00 8445.ca1&s... (15 March 2009).

58. "Journalism Evolving, Not Dying: Science Author," *Breitbart.com*, 14 March 2009, http://www.breitbart.com/print.php?id=CNG.99c75c577fa1a698c1689adb0e00 8445.ca1&s... (15 March 2009).

59. Morley Winograd and Michael D. Hais, *Millennial Makeover: MySpace, You-Tube, and the Future of American Politics* (New Brunswick, NJ: Rutgers University Press, 2008).

60. "Americans Spend Eight Hours a Day on Screens," *Breitbart.com*, 27 March 2009, http://www.breitbart.com/article.php?id=CNG.92e661444313b232e8931de00c 29c73b.431... (28 March 2009).

61. "Americans Spend Eight Hours a Day on Screens," *Breitbart.com*, 27 March 2009, http://www.breitbart.com/article.php?id=CNG.92e661444313b232e8931de00c 29c73b.431... (28 March 2009).

62. Douglas Ahlers and John Hessen, "Traditional Media in the Digital Age: Data About News Habits and Advertiser Spending Lead to a Reassessment of Media's Prospects and Possibilities, *Nieman Reports* (Fall 2005): 66.

63. "'Engage or Die': 3i's Latest Report on the Future of the Media Sector given Exponential Growth in User Generated Content," *M2PressWIRE*, 1 November 2007, http://web.ebscohost.com/ehost/delivery?vid=6&hid=112&sid=1105a8e1-a307-4... (8 February 2008).

64. Abbey Klaassen, "Are eBay and Others the New New Media?" *Advertising Age* 78, no. 37 (17 September 2007): 3–68, 2p, 1c.

65. A. C. Croft, "Emergence of 'New' Media Moves PR Agencies in New Directions: Competitive Pressure Threatens Agencies' Livelihood," *Public Relations Quarterly* 52, no. 1 (1 April 2008): 16–20.

66. Croft, "Emergence of 'New' Media Moves PR Agencies in New Directions: Competitive Pressure Threatens Agencies' Livelihood," *Public Relations Quarterly* 52, no. 1 (1 April 2008): 16–20.

67. Ben Grossman. 2009. "Confessions of a Twitter Cynic." *Broadcasting & Cable*, 11 May, 8.

68. Alex Weprin. 2009. "Much A-Twitter about Something." *Broadcasting & Cable*, 11 May, 12–13.

69. Scott Davis, "Don't Be Afraid to Plunge into Emerging Media," *Advertising Age* 78, no. 31 (6 August 2007): 15.

70. Davis, "Don't Be Afraid to Plunge into Emerging Media," *Advertising Age* 78, no. 31 (6 August 2007): 15.

71. Douglas Ahlers and John Hessen, "Traditional Media in the Digital Age: Data About News Habits and Advertiser Spending Lead to a Reassessment of Media's Prospects and Possibilities," *Nieman Reports* (Fall 2005): 65–68.

72. Ahlers and Hessen, "Traditional Media in the Digital Age: Data About News Habits and Advertiser Spending Lead to a Reassessment of Media's Prospects and Possibilities," *Nieman Reports* (Fall 2005): 66.

73. Rupert Steiner, "Old Media Will Finally Learn to Love the Net," *Sunday Business* (United Kingdom), 23 December 2006, http://web.ebscohost.com/ehost/delivery ?vid=8&hid=112&sid=1105a8e1-a307-4... (8 February 2008).

2

Media Management

The Changing Media Industry and Adaptability

Mary Jackson Pitts and Lily Zeng

"Pictures in our heads" are created by an increasing number of information and entertainment outlets that reach us by air and wire. Broadcasters are equipped to compete in this environment using an arsenal of delivery mechanisms. Armed with video and audio, radio and television broadcasters can use their expertise to develop content, and users will decide what, when, where, and how to consume the content of their selection. Broadcast content is ripe for use in every format of media delivery because of the ease of transforming broadcast material into every type of media-on-demand format. Bountiful bandwidth and high quality equipment provide this generation of broadcasters with substantial advantages over past broadcasters. The digital generation of broadcasters has what "feels like" an infinite number of channel outlets.

Creating mobile content takes the broadcaster wherever the consumer wants to go and allows the broadcaster the opportunity to reach an extended audience past the radio and TV bandwidth. With simple software, content created for the airwaves finds its way onto mobile devices. In addition, media-on-demand users are generally more motivated than channel surfers, creating great opportunities for niche marketing and advertising. The potential for economic growth, however, raises historical challenges for media management. More detailed audience data need to be collected, analyzed and used to decide which niche audience groups a media organization serves. Media producers need to be trained to understand different expectations and usage patterns by different audience groups. The traditional thinking of mass delivery will be abandoned to ensure the recognition of the importance of the audience as individuals who want MY news and/or MY entertainment. In this chapter, we explore how digital broadcast changes how and when audiences consume what content. Historically, broadcasting has always been on edge with its audience, constantly changing to create content that attracts the consumer. Broadcast managers must learn from the past to create programming that benefits the consumer in the future.

History

KDKA's broadcast as a licensed commercial radio station on November 2, 1920, was the first of many advances in broadcasting. Radio stations in the 1920s worked hard to provide content that was both useful and entertaining. The advent of radio networks helped the stations supply information about politics and provide radio entertainment. Radio managers moved into the 1930s seeking ways to monetize their efforts. Advertising-supported content allowed for "free" entertainment and the audiences grew exponentially in the 1930s and 1940s.[1] Leading up to World War II, radio flourished as people sought information about the rising conflict in Europe, and television was introduced at the World's Fair as the 1930s drew to a close. With all efforts focused on the war, radio remained the main information source through World War II. But as the war drew to a close, the focus began to shift to television and the type of content that would attract viewers to the screen. At the start of 1952, there were about 20 million homes that had a television set but as the decade ended more than 45 million homes had a television set.[2]

Television executives were fast learners as the industry evolved in the 1950s and 1960s. Programming strategies that worked for radio worked equally well for television. The radio network model was used in television with ABC, CBS and NBC emerging as the primary television networks. Television network executives quickly provided entertainment programming that allowed television to become the primary source of information and entertainment for most Americans. They often used programming such as soap operas, westerns, quiz shows, and variety shows to capture the television audience. Television audiences exploded in the 1960s as major events in U.S. history drew people to the television screen. Images enough to last a lifetime were flashed across the screen, first in black and white and then in color. This was the coming of age of television journalism in the 1960s and news proved to be local television's most saleable content (e.g., real-life coverage of presidential debates, the space race, the death of a president, the civil rights movement and the Vietnam War). Television network entertainment programming was an even greater draw. As more and more viewers consumed television content, there was fear that the television networks wielded too much power and the Federal Communication Commission (FCC) stepped in to limit this power in the early 1970s.

Television networks were kept from having significant financial interest in the syndication market and were only allowed to broadcast three hours of prime-time network programming a night. These actions were taken to encourage local television markets to provide more locally originated programming.[3] Instead of creating locally produced programming, many local affiliates chose to use syndicated programs with game shows and talk shows as typical choices. The syndication industry revenue and the television audience increased and by 1980 there were television sets in 97.9 percent of all U.S. households.[4]

Satellite delivery gave television stations with limited coverage area the opportunity to expand into new markets by distributing station signals to cable systems across the country. The stations found favor among audiences because of their diverse programming content and quickly rose in prominence when the FCC moved to allow an open entry policy for cable.[5] This meant that local affiliates would be touched by added competition. The television audience as a whole increased but the audience's attention was divided among more choices and loyalty to one or two stations was changed. This forced local television managers to evaluate programming content and make decisions that would stabilize their position among audiences.

Radio struggled as television pulled the audience away but radio found new life with the FM signal during the 1970s and 1980s. The clarity and quality of the FM signal and radio programmers' willingness to move away from the traditional formats found on AM stations positioned FM stations as the leader in the radio industry. Managers changed the way they thought about their audiences and turned what were once money-losing FM stations into multi-million-dollar businesses by adding new content and using syndicated music programming to attract audiences. Albarran suggests that "the radio industry recognized that to survive, it had to rely on the local market for economic support."[6]

Market shares for network television dropped in the 1980s but the local station affiliate revenue increased as managers made decisions to increase the amount of local news programs. Stations added a second evening newscast, often prior to the network news programs. Some local stations turned their 6:00 p.m. newscast into an hour-long newscast. When television ownership restrictions were lifted in the mid-1980s this allowed corporate ownership of network affiliate stations, and, for the most part, gave owners the ability to pool resources to provide for more efficient management of human and property resources. One television player to emerge from the change in ownership rules was Fox Broadcasting. By 1987, Fox had become "the 'official' fourth" television network. Fox started with limited programming and by the mid-1990s provided programming similar to that of the other three networks. A move by New World Communication to move all 12 of its stations to become Fox network affiliates threw the television network industry into great upheaval in May 1994. While hurting the networks, this move helped the local affiliates because individual stations were able to negotiate network compensation packages that were more favorable toward the affiliate.[7]

Broadcast companies became mega media companies as the FCC increased the cap on the number of radio and television stations a group could own. Viacom bought Paramount, Disney bought ABC, Time Warner took over Turner Broadcasting, and Westinghouse purchased CBS. This brought on more competition for the four networks—ABC, NBC, CBS, and Fox—when WB and United Paramount Networks formed.[8] When the Telecommunication Act of 1996 was passed, national radio ownership limits were eliminated, allowing companies to own an unlimited number of radio stations but with some restrictions on the number of local radio stations and television ownership was tied to the cap on the total national audience reach. From this change emerged new owners for CBS and Media One, while Clear Channel wound up owning more than 800 radio stations. Running parallel to these changes in the television industry were mergers in the telecommunications industry, which meant changes in how cable companies operated and how traditional phone companies were allowed to get into the business of disseminating content. While this was occurring in the 1990s, some broadcasters were taking the first step toward offering digital signals. The first digital television station went on the air in 1996 in Raleigh, North Carolina, and the Federal Communication Commission assigned digital channels to all analog television stations in 1997.

By 2000, phone and cable companies were able to provide more services similar to that of a traditional broadcaster and with digital technology more channels were offered to consumers. This created an even tighter squeeze on the traditional broadcaster, who seemed willing to allow the marketplace to decide when or if it would broadcast using digital technology. Media consumers emerged as victors because the choices for information and entertainment further increased. The broadcasters, for fear of losing audiences, would not convert to digital-only signals because too many of the consumers did not have the capability of receiving the digital signal. In 2004, the FCC relaxed ownership rules again to accommodate ownership of more television stations in local markets. In 2005, ABC provided online availability of some of its television programming. The industry suggested a repurposing of broadcast content. Miller suggested that everyone recognized by 2006 that "the Web would eventually become a TV medium at least the equal of broadcasting, cable, and satellite."[9]

While government regulation of broadcasters has always been a motivating factor in broadcast programming and operation strategies, the biggest single event in broadcasting history happened in June 2009. Television broadcasters, almost 1,400, went all digital. A small number of radio broadcasters use digital but most still use the analog signal. This move was 13 years in the making.[10] Only 2.8 million of the 114.5 million U.S. television households were not prepared for the June 2009 transition. This number was helped by the FCC's $1.5 billion coupon program that gave users the money to purchase digital converter boxes.

After years of the government stepping in to regulate and to deregulate broadcasting, the horizon for television and even radio is brightened by the increase in the bandwidth and ultimately the increase in content choices that local broadcast stations can offer. The transition allows broadcasters the ability to multicast. Multicasting enables broadcasters to broadcast multiple channels at the same time. So instead of one TV channel 8, a station might have channels, 8.1., 8.2, etc. (Radio and Television Broadcast Rules 47 CFR Part 73, 2009). Likewise, radio can offer multicasting as well. For radio, multicasting is often called HD radio. However, the radio stations are not required to switch to digital only. Radio stations can broadcast in analog and digital simultaneously. What radio and television stations get out of digital is the ability to multicast and the opportunity to improve the quality of their transmission signal. The 1,378 commercially licensed television stations and the 11,213 commercial radio stations should use this opportunity to create content that reaches the "me" users, which in turn will help revitalize television and radio broadcasting.[11] Doing so will position broadcasters to maximize profits with diverse audiences. Media-on-demand content is the direction in which broadcasters should focus their attention.

This opportunity was tempered by the worst economic situation since the Great Depression. In 2009, unemployment in the U.S. climbed to more than 10 percent, with some expecting it to go higher. Without people working, consumer spending went down and consumer confidence waned. Ad dollars that have typically come from the automotive industry and banking dropped significantly. Erik Kolb and Tuna Amobi suggest a significant pullback in local advertising dollars for media, while also seeing pullbacks at the national level.[12] In 2002, $61.54 billion was collected by broadcast companies. That revenue rose to more than 68.8 billion in 2006 but began to drop in 2007 and further declined to 59 billion in 2009, an almost 14 percent decline in revenue over a three-year period.[13] In addition, television stations have spent millions preparing for the digital television (DTV) transition increasing their debt load and then facing increased competition from Internet options, satellite radio, iPods, and other mobile devices. To create a bright spot, broadcasters must recognize the opportunities created by DTV, HD radio, the Internet, and mobile on demand media.

Data show that radio and television station Internet revenue is up. BIA advisory services and the National Association of Broadcasters (NAB) found of the $11 billion spent on local online advertising, 7.3 percent ($805 million) was generated by broadcasters. BIA is predicting broadcast revenue to increase by 18.6 percent in the next five years as broadcasters begin to further promote their Internet content on handheld devices.[14] Local advertisers increased their online ad spending by 46 percent in 2008 and the number was expected to increase in 2009. Erik Kolb and Tuna Amobi suggest, "if broadcasters can effectively leverage their online platforms, the area seems ripe with cross-promotion opportunities."[15] Broadcasters would benefit most if they were to adopt a single mobile DTV standard. Some predict that for each month this is delayed broadcasters will lose 50 million dollars.[16] DTV allows stations to expand their horizons by having more than one channel outlet and broadcast executives must use the multiple platforms to distribute their content. These platforms create value-added content that cannot be dismissed. Each platform becomes an opportunity to create more useful content for the "me" user.

Because of significant changes in technology, globalization, regulatory reform and demographic change, broadcasters must develop "innovative strategy and financial flexibility" to survive the competitive marketplace. Kolb and Amobi suggest broadcasters must "devise new growth strategies to augment [their] core advertising revenue model."[17] Development of these new growth strategies will have major implications for management and economic philosophies. The most significant change will come when broadcasters emphasize the user of media.

Economic and Management Implications

From Mass to Niche Audience

Broadcast, cable, and satellite industries have distinct ultimate goals due to the different business models in which they operate. For instance, the main revenue source for terrestrial television and radio broadcasters is advertising. Therefore, the goal of these companies is to boost the ratings of the stations and the networks these stations are affiliated with so that advertising can be sold at a higher rate. For cable providers and satellite television and radio broadcasters, however, the main goal is to attract new subscribers while retaining current users. After all, commercial media companies are economic entities/institutions and thus are driven by profit.

Regardless of the difference in their business models and therefore their ultimate goals, the route to success for all television and radio firms is to provide programming that appeals to a targeted audience. The rule of "lowest common denominator" proved to be a success in a marketplace of scarce viewing options. Through the offering of a little something for everybody, the three broadcast networks secured over 90 percent of prime-time television viewership in the 1970s.[18] Members of a targeted audience share their commonality in demographics such as age, gender, and socioeconomic status (SES), creating mass audiences in the traditional sense.

Demographics of media audiences are important variables in determining advertising rates for media businesses that rely on advertising. Both radio advertising rates (measured in cost per point or CPP) and television advertising rates (measured in cost per thousand viewers or CPM) reflect stations' ability to attract targeted demographic audiences, among other factors.[19] Attention of audiences becomes the invisible but measurable product marketed by media companies. Media companies could sell the audience attention they attract to buyers at advertising agencies.

While the radio industry has long been operating on the basis of localized target audiences, the television industry did not see an explosive popularity of niche programming to specific audiences until the 1990s.[20] Since then, there has been "continued erosion of audience share due to fragmentation."[21] As of 2009, there are over 500 cable channels operating in the United States.[22]

On both the economic and management levels, further fragmentation of audiences poses challenges to the rule of "lowest common denominator." In response, channels choose to specialize in particular types of content. Stations "narrowcast" whatever types of content they believe will attract a desired demographic, and seek to send the attention of the audience group to advertisers who are willing to pay for access to the demographic. However, demographics such as age and income alone are no longer sufficient in studying the fragmented traditional mass audience. Narrowcasters should study their niche audiences based on psychographic (lifestyle) combinations of interests and attitudes. A well-based understanding of the audience and efficient measurement of such audience attention will transform narrowcasters into successful nichecasters in an increasingly competitive market of option abundance.

Content Diversity

Multicasting opens programming choices for television and radio. Multicasting gives local stations the ability to program content that can reach diverse audiences. It is only right to remind broadcasters that the FCC handed broadcasters this gift of multicasting at no cost. In 2000, former FCC chairman William E. Kennard warned broadcasters to remember the "public interest" broadcast mantra especially in light of the fact that broadcasters had been given more than $70 billion in spectrum space when DTV was allocated in 1997. He reminded broadcasters that "this gift to broadcasters stands in stark contrast to other users of the spectrum—like wireless providers—who have paid billions for licenses to use the airwaves."[23] As suggested by Kennard, "as broadcasters reap many billions of dollars from their use of the airwaves, they must also use the airwaves to serve the public interest."[24] Broadcasters should provide local content to meet the public interest charge. People can get information from millions of sources but getting local information takes a commitment at the community level of operation. Managers that begin to recognize the importance of creating local content down to the lowest level of news coverage will see viewership increase. Leveraging "original local programming, including staples such as news, weather," will attract consumers that advertisers want to reach. It comes down to monetizing local content.[25]

As broadband Internet continues its penetration into an average American household, online delivery channels are able to provide within easy reach content no longer limited strictly by space, time, and bandwidth. For traditional television and radio firms, opportunities emerge to develop content that appeals to smaller segments of audiences who have specialized needs for information. Localized content holds value for the dual role of media firms: to attract local audiences as well as local advertisers who are interested in buying attention from those audiences.

As media companies try to address the special needs of smaller audience groups, there is a dramatic increase in content diversity and therefore in the total amount of original content that should be produced. One of the key economic characteristics of media products is high production costs of the first copy, and minimum costs for additional copies and distribution.[26] A lower demand for copies of diversified content means a higher production price per copy. Moreover, the higher the costs of the first copy production, the higher price per copy of the product. Therefore, diversity is "expensive to achieve" unless low-cost production strategies can be implemented.[27]

This increase in demand for diversity content creates new challenges for media managers. On one hand, few, if any, television or radio firms can afford the production of such an increased amount of content. Media managers must seek new opportunities to reduce production costs. On the other hand, the economic success of a media firm depends heavily on the creation and retention of an audience that can generate sufficient revenue. Any addition to the audience base, regardless of the size, will incur hardly any extra cost but can increase the revenue. Managers who can expand their target audiences will no doubt place themselves in a better economic position in the highly competitive media market. However, continued segmentation often means shrinking audiences. Therefore, it is extremely difficult for a nichecaster to develop a large audience, not to mention a mass audience. Despite the inapplicability of economies of scale, managers may find it an effective way to boost their revenue if they can enhance the engagement of their audience.

Cost Reduction

One way to reduce production costs is to take advantage of the existing news crew and prepare them for the new responsibilities in the digital age. While production teams at media firms traditionally consist of professionals with specialized skills, what media firms need now are professionals with a broader set of skills. Backpack/video journalists, or individuals who can function in a wide variety of roles, are in high demand. Media managers, therefore, face the challenge of training existing crews so each production professional can serve as a reporter, photographer, and videographer, among many other roles.

Thanks to technological advancement, media equipment is becoming smaller, lighter, less expensive, more user-friendly, and multifunctional. Even during hard economic situations like the current one, many media firms can afford a light-weight "backpack" of professional quality equipment for individual production professionals. In addition, it is not impossible for a professional with a specialty in one area to learn convergence skills. Radio professionals, for example, can learn to write for the Web, shoot pictures, and even record and edit video for their website. Even when working within their traditional specialty area, media professionals often should adjust their content according to the audience characteristics of the specific medium. For example, most television reporters no longer directly shovelcast their newscast on their media website, because online audiences differ from television audiences in their expectation and use patterns.

One recent initiative of equipping media professionals with the latest technology to help them function in multiple roles happened at KOMO, a Seattle, Washington, television station owned by Fisher Communications.[28] Instead of carrying a heavy laptop around, each photographer and reporter received from the station an iPhone, which is used to take pictures and send them back to the station with brief text description of the unfolding story. It proved to be an economically manageable initiative, since no additional employment is involved. However, the success was based on a lot of explanation and coaching before the reporters incorporated the technology in their daily reporting jobs. Media managers, therefore, face the challenge of educating and training the crew how to use new technology to fulfill the responsibilities they are used to.

The multicasting opportunity as a result of the digital TV conversion presents a dilemma for television managers. On one hand, they often find it difficult to rely solely on their existing crew to produce enough content for the additional channels they have acquired. In addition, as the Internet continues to serve as a favorite source of media content for a wide range of demographics, television websites must provide content in an amount and variety beyond the capabilities of their crew. For example, as a story develops, the online audience may expect up-to-the-minute updates, or "twitcast," a Twitter version of a live event as it unfolds. On the other hand, new employment will incur additional costs, which have to be offset by even higher revenue in an extremely competitive and deteriorating market daunted by job cuts. In much the same vein, visitors to a radio website expect more than just streaming audio. They also want detailed background information about the music, pictures and biography of the musician, and even videos of relevant concerts. Most important, television and radio audience members want to establish a personal relationship with the anchor or DJ of their favorite program. They want to read the anchors'/DJs' bios, learn about their personal lives, friend them on Facebook or MySpace, and follow them on Twitter.

In the midst of multicasting, there comes multitasking, where the employee is over-burdened with providing content for old and new platforms of delivery. Therefore, it is not enough to simply equip every crew member with an iPhone and train them how to use it to do their jobs. Media managers must seek new strategies to relieve the burden of creating multi-platform content. One important solution for low-cost production is user-generated content. Recent years have witnessed numerous examples of users contributing to the normal media production process, especially in the case of unexpected events (e.g., witnesses posting videos of the Virginia Tech campus shooting in 2008). Major television stations have made user contribution a standard form of content production. For example, CNN uses "iReport," which is listed side by side with its main news categories on the website. Similarly, Fox News offers a "U-report" capability on the top right corner of the first page of every section. Local television stations such as KARK, in Little Rock, Arkansas, invite the general audience to contribute pictures and video through "MyMedia." The advancement of consumer technology, particularly mobile phones with photo, video, and Internet connection capabilities, makes it easy for users to capture and share stories that are beyond the reach of news professionals, particularly hyperlocal news that serves a small neighborhood. Television stations that have over the decades provided "Eyewitness News" should adopt the model of "iWitness News," where "i" is the active audience reporter seeking to participate in the content production process.

One dilemma media managers face in the use of user-generated content is the balance between user contribution and newsroom standards. Because general users have not gone through journalism school training, using the content they produce may mean a sacrifice in journalistic standards. There is no single correct solution to this problem. However, most journalists are concerned about the quality of user-generated content. The Seattle-based TV station KOMO, for example, does not allow user-generated content to be directly posted on its website without screening by designated staff.

User-generated content can prove to be an effective way to produce content that appeals to small groups of local audiences and, with appropriate marketing strategies, can help produce economic success. KOMO innovated with a hyperlocal platform that connects "neighborhood content, local bloggers and user-generated material."[29] Using geotagging technology provided by a partner data company, KOMO geographically targets advertising to specific neighborhoods. The success of the new strategy is demonstrated by the fact that the platform has attracted hundreds of new clients, many of which are small businesses that either could not afford advertising on television or could not find a suitable audience group.

Revenue Increase

While the economic viability of media firms depends on low cost, it is even more important to boost revenue through improving advertising ratings and generating new revenue from sources such as multi-platform distribution of repurposed content. Media managers, therefore, face a series of revenue opportunities as well as challenges.

Repurposing Content for Multi-platform Delivery

There has been fear that as technologies advance and media options multiply, television viewers, particularly the younger generations, might move to the Internet and even mobile devices for content that has traditionally been available only on television. The fear seems justified since we only have 24 hours a day and the more time we spend on one medium, the less time we have left for the other media, as predicted by McCombs and Eyal's principle of relative constancy.[30] However, it was found that different media are used as complements, not substitutes.[31] More recently, audience data suggest that television, computer and mobile devices each serve unique functions. Instead of migrating to the Internet, and more recently, mobile devices, television viewers actually watch more television.[32] For example, during the first quarter of 2009, the average American watched 153 hours of television every month at home, and the 131 million Americans who watch online video watched on average about 3 hours of online video every month at home and work.[33] The increase in use across different media platforms is largely made possible through multitasking. For example, a large proportion of the television viewers are on the Internet while watching television, making them simultaneous users of the two media. Data from the Nielsen TV/Internet Convergence Panel reveal that over half of the 3,000 people on the panel are simultaneous users of television and the Internet.[34] More important, simultaneous users watch 14 percent more TV a day and use the Internet 61 percent more than the average consumer. "The simultaneous usage phenomenon presents new marketing opportunities: the unique strengths of each medium can be leveraged to allow consumers to be reached—and allow them to reach back—in ways that they choose themselves."[35]

The broad use of mobile phones and broadband Internet connection (accessible to nearly 3 out of 4 Americans), plus the advanced capabilities of mobile devices to handle multimedia content and Internet connection, help establish mobile devices as "the next frontier for media opportunities."[36] In 2009, 9 percent of those 12 years and above watched TV clips on a cell phone,[37] and the amount of usage is predicted to grow as cheaper mobile phones are configured to handle complicated data transmission in mobile video viewing and service costs decrease to a level acceptable to more users.

Therefore, television and radio firms should react quickly to this user trend and expand their territory to the mobile platform through repurposing their content. Repurposing content produced for one medium and using it for another usually involves technical transformation, which can be accomplished automatically or involves a single additional click of a button, thanks again to the development of technology in recent years. The role of media managers, in most cases, is to reallocate resources to ensure that mobile delivery as an option is available for all content that may attract additional users, which means a possible source of revenue.

Because broadcast radio and local television are geographically bounded industries, economics of scale has limited applicability in an age of proliferated new channels and thus continuing audience segmentation. Multi-platform delivery, however, helps make up for what is lost in one city in one medium by attracting small audience groups in other cities (sometimes in a different country) in another medium. In this sense, repurposing content for multi-platform delivery can overcome the weaknesses of economics of scale through economics of scope.

Interactivity

Interactivity happens when the audience is involved in the production of media content, through direct modification, feedback, or even the creation of original content. Interactivity is an essential feature of online media content.[38] In the Web 1.0 days, hyperlinks on a media website took the audience to other parts of the same site or even a site hosted in another country for additional information of relevance. The audience could also e-mail the reporter or producer with comments. Web 2.0 applications such as blogs, social networking, and more recently, Twitter and mixx, allow the audience to more fully interact with the original content provider, share comments of content with friends, provide updates, and even vote on the popularity of a piece of content. Therefore, content created by the production team does not end the moment it is published by a media firm. Audience members take the topic and continue the production process through interacting with others ranging from the original creator to other individuals interested in the topic. Comments or notes that audience members post on anchors' Facebook, MySpace or Twitter accounts can end up being aired in the actual programs.

The continuation of the production process by the audience creates unprecedented opportunities for media firms. A high degree of audience interactivity suggests a high level of audience attention and involvement, which creates value for advertisers who want to appeal to such demographics and psychographics. However, the existing measurement of audience attention fails to take into account the amount of involvement from certain audiences. Because the variance in audience involvement is related to selective exposure and cognition, which may lead to variance in attitude and behavior, audience involvement is a more important variable than attention alone for advertisers. Media firms should work with ratings companies to create and test new measures to establish advertising rates in the interactive media market. Most recently, fourteen companies ranging from broadcasters to advertisers coalesced to form the Coalition for Innovative Media Measurement.[39] Effective measures, once developed, should work in the benefit of both media and advertising firms since the measures will be based on a better understanding of the audience. Media managers should seek to develop localized audience measurements. These measurements can be implemented with easy online survey software.

Two other aspects of interactivity that can lead to economic success of media firms are cross-promotion and direct selling of products. The turn of the twenty-first century saw a shift of the focus of marketing strategies from branding to cross-promotion.[40] Media audiences are accustomed to messages on television such as "visit our Web site at . . ." More recently, media companies started doing a more effective job at sending Web visitors to the primary medium by encouraging them to "tune in . . ." or to watch the next episode of a show on television.

Another direction some successful media businesses are going is the use of external links (i.e., linking to other sources for information that they themselves do not provide). It is not surprising that media managers will feel a sense of fear to send their own audiences to the competitor. However, the rationale behind such a seemingly risky practice is that "you can't serve ALL of the needs of your customer yourself, then the best that you should do is to be the FIRST source of information for your audience."[41] The fear and risk comes from a loss of control. Several online-only media companies, such as CNET.com and News.com, have been successful in implementing this strategy. This "become the FIRST source for your audience" strategy may pay off for managers who want to see a more active role for their firms on the Internet. The challenge is to deal with possible changes in streams of revenue when the audience is directed to a likely competitor.

In some cases, advertising alone does not generate sufficient revenue for media firms. Direct selling of merchandise or advertised goods can become a possible revenue source. Media companies that are accustomed to producing content to attract audiences can allow interested audience members to purchase a media product or an advertised product or service directly from the computer, a mobile device, or on the television screen (in the near future). This option will extend media firms beyond their traditional boundaries as content providers and transform them into service providers. Media managers need to further monetize their local online advertising space through direct purchase of the advertised products or service. Partnership opportunities and even ready-made affiliate schemes are already available.[42] The direct purchase capability on media websites allows media managers and advertisers to assess the effectiveness of advertising. Media managers should develop formulas to establish advertising rates. These formulas should include data on psychographics and individual and collective purchasing behavior, bringing advertising to the level of personalized advertising, or iAd. Media managers should also explore the impact of direct selling on the revenue streams of television and radio platforms.

Conclusion

Broadcast managers have been passive participants in their own history, allowing policy changes and technology advances to dictate their management styles instead of listening to their audiences. The audience may have been heard but their comments and criticism did not affect how media content is produced and programmed. With the exception of ratings months, the audience was seldom considered. Now with technological advances, the audience has the opportunity to provide almost instantaneous comment and criticism. Those who decide to provide comment are typically active participants who take the initiative to switch channels or to cross media platforms in search of information that appeals to their specialized needs.

The manager's goal should be to empower the audience member to share information to create a sense of audience involvement that turns into a community investment, yielding a station profit. Managers should rally their own employees to recognize the power of the "iReporter," seeing them not as intruders but as helpers in the newsgathering process, giving the station the power to report down to the street level just as the Google map takes them there. Just as the iReporter sends the station to the street level in content production, so should managers "geotag" neighborhood content to create iAds. The creative manager will be as active as his/her active audience members, using specialized measurement to monetize every audience member's investment in every platform of the local media. They will take their future into their own hands.

Notes

1. Mark K. Miller, "A Brief History of Broadcasting and Cable," *Broadcasting and Cable Yearbook 2009* (New Providence, NJ: ProQuest LLC, 2009).

2. Miller, "Brief History."

3. Alan B. Albarran, *Media Economics: Understanding Markets, Industries and Concepts*, 2nd ed. (Iowa State Press, 2002).

4. Mark K. Miller, "A Chronology of the Electronic Media," *Broadcasting and Cable Yearbook 2009* (New Providence, NJ: ProQuest LLC, 2009).

5. Miller, "Brief History."

6. Albarran, *Media Economics*, 61.

7. Albarran, *Media Economics*.

8. Miller, "Brief History."

9. Miller, "Brief History," A–19.

10. Erik B. Kolb and Tuna N. Amobi, "Broadcasting, Cable & Satellite," *Standard & Poor's Industry Surveys* (New York: Standard and Poor's Equity Research Services, 2009), 1–33.

11. Kolb and Amobi, "Broadcasting, Cable & Satellite."

12. Kolb and Amobi, "Broadcasting, Cable & Satellite."

13. Kolb and Amobi, "Broadcasting, Cable & Satellite."

14. Glen Dickson, "Findings Suggest Quick Rollout is Necessary to Maximize Revenue," *Broadcasting & Cable* (5 Feb. 2008), http://www.broadcastingcable.com/

article/print/95522-OMVC_Backs_NAB_BIA_Study_on_Mobile_DTV.php (1 Nov. 2009).

15. Kolb and Amobi, "Broadcasting, Cable & Satellite," 8.

16. Glen Dickson, "Findings Suggest Quick Rollout Is Necessary to Maximize Revenue," *Broadcasting & Cable* (5 Feb. 2008), http://www.broadcastingcable.com/article/print/95522-OMVC_Backs_NAB_BIA_Study_on_Mobile_DTV.php (1 Nov. 2009).

17. Kolb and Amobi, "Broadcasting, Cable & Satellite."

18. Shuler Veronis and Associates, *Communications Industry Forecast* (New York: Author, 1994).

19. Kolb and Amobi, "Broadcasting, Cable & Satellite."

20. Kolb and Amobi, "Broadcasting, Cable & Satellite."

21. Dan Shaver and Mary Alice Shaver, "Directions for Media Management Research in the 21st Century," in *Handbook of Media Management and Economics,* ed. Alan B. Albarran, Sylvia M. Chan-Olmsted, and Michael O. Wirth (Lawrence Erlbaum Associates, 2006), 639–54. p. 648.

22. Kolb and Amobi, "Broadcasting, Cable & Satellite."

23. Kennard, William E., "What Does $70 Billion Buy You Anyway? Rethinking Public Interest Requirements at the Dawn of the Digital Age," A speech given at the Museum of Television and Radio. New York, NY. http://www.fcc.gov/Speeches/Kennard/2000/spwek023.html (October 29, 2009), para 22.

24. Kennard, "What Does $70 Billion Buy," para 24.

25. Kolb and Amobi, "Broadcasting, Cable & Satellite," p. 14.

26. Waterman, David, "The Economics of Media Programming," in *Handbook of Media Management and Economics,* ed. Alan B. Albarran, Sylvia M. Chan-Olmsted, and Michael O. Wirth (Lawrence Erlbaum Associates, 2006), 387–416.

27. Waterman, "The Economics of Media Programming," 390.

28. Tom Petner, "Hyperlocal More Than a Slogan to Fisher," http://www.tvnewscheck.com/articles/2009/10/19/daily.11/ (20 Oct. 2009).

29. Petner, "Hyperlocal," para. 6.

30. Maxwell E. McCombs and Chaim H. Eyal, "Spending on Mass Media," *Journal of Communication 30,* no. 1 (1980): 153–58.

31. Brian Carroll, "Newspaper Readership v. News Emails: Testing the Principle of Relative Constancy," *Convergence 8* (2002): 78-96.

32. Horst Stipp, "Convergence Now?" *International Journal on Media Management 1* (1999): 10–13.

33. Nielsenwire, "Americans Watching More TV than Ever; Web and Mobile Video Up Too," http://blog.nielsen.com/nielsenwire/online_mobile/americans-watching-more-tv-than-ever/ (28 Sept. 2009).

34. Nielsenwire, "Multitasking at Home: Simultaneous Use of Media Grows." http://blog.nielsen.com/nielsenwire/online_mobile/multitasking-at-home-simultaneous-use-of-media-grows/ (28 Sept. 2009).

35. Nielsenwire, "Multitasking."

36. Arbitron, "The Infinite Dial 2009," http://www.arbitron.com/study/digital_radio_study.asp (15 Oct. 2009), 64.

37. Arbitron, "The Infinite Dial."

38. E.g., Joshua D. Atkinson, "Towards a Model of Interactivity in Alternative Media: A Multilevel Analysis of Audiences and Producers in a New Social Movement Network," *Mass Communication and Society 11* (2008): 227-47; Danielle Endres and Barbara Warnick, "Text-based Interactivity in Candidate Campaign Web Sites: A Case Study from the 2002 Elections," *Western Journal of Communications 68* (2004): 322–43.

39. Eric Deggins, "A New Way to Measure TV Audiences Besides Nielsen? Stay Tuned," http://www.tampabay.com/features/media/a-new-way-to-measure-tv-audiences-besides-nielsen-stay-tuned/1035213# (1 Oct. 2009).

40. Susan Tyler Eastman, Douglas A. Ferguson, and Robert A. Klein, "Promoting the Media: Scope and Goals," in *Media Promotion & Marketing for Broadcasting, Cable & the Internet,* ed. Susan Tyler Eastman, Douglas A. Ferguson, and Robert A. Klein (Focal Press, 2006), 1–30.

41. Charlene Li, "The Changing Media Business Model," http://blogs.forrester.com/groundswell/2006/05/the_changing_me.html (20 Sept. 2009), para 11.

42. Paul Bradshaw, "Making Money from Journalism: New Media Business Models (A Model for the 21st Century Newsroom pt5), *Online Journalism Blog* 2008, http://onlinejournalismblog.com/2008/01/28/making-money-from-journalism-new-media-business-models-a-model-for-the-21st-century-newsroom-pt5/ (20 Sept. 2009).

3

DVRs and the Empowered Audience

A Transformative New Media Technology Takes Off

James R. Walker and Robert Bellamy

T HE TELEVISION INDUSTRY HAS BEEN SHAPED by two generations of change.[1] In the first generation, it was dominated by two, and later three, major networks that generated similar types of advertiser-supported content, while audiences selected the "least objectionable programming" offered by the broadcast networks. The second generation of television saw the diffusion of cable, VCRs and inexpensive receivers with remote control devices (RCDs). This brought dramatic changes to the television industry and the viewing environment. Audiences gradually changed from families watching the big three broadcast networks to individuals choosing from an ever-expanding array of video outlets. However, except for a small amount of VCR time-shifting, viewers still selected from the programs available at a particular time, and the television industry was still supported primarily by advertising presented in commercial announcements. Although the television user could still be conceptualized as a "viewer" of television, the second generation was the real beginning of the viewer as "user," an active participant in manipulating the available content.

The development and increasing penetration of digital video recorders (DVRs) is one of many forces (e.g., the Internet, mobile video) pushing television's evolution into a third generation. Consumers who have fewer common television experiences will challenge time-tested assumptions about advertising, programming, and even the definition of the audience. In this third generation, the idea of the consumer as "viewer," with its connotations of passivity and target, largely will yield to conceptions of the audience as "users" of the medium. Viewer/users are likely to pay for specific programs

and program packages via some variation of an à la carte pay-per-view (PPV)/
video-on-demand (VOD) basis rather than for an array of television networks
many of which they do not watch. Or they may simply download digitized
content into their iPods for later viewing on either the small (MP3 player) or
large (LCD, plasma) screen. Distribution will come in wired (cable TV, wired
broadband), wireless (satellite, cellular, wireless broadband, broadcast), and
physical (DVD, Blu-Ray, digital drives) forms.

More advertising will shift from spot announcements to program place-
ments and other forms of zap-proof marketing/advertising. Indeed, this
process is well underway, with *BusinessWeek* noting the rapid decline of mass
market and spot advertising even before the mid-point of the new millennium's first decade.[2] As spot advertising declines, the seam that distinguishes
commercial from program will become less and less visible. The advertising-
supported model of "free" broadcasting, so popular with consumers mired in
the Great Depression of the 1930s, has given way to more complex financial
models that draw revenues from many additional streams, including enhanced websites, Internet downloads, DVD and mobile video sales, retransmission fees from cable/satellite providers, and product placement.

DVR Diffusion

The penetration of DVRs into U.S. households has progressed at a more lei-
surely pace than first predicted after their appearance in 1997.[3] As of March
2009, Nielsen reported that 30.6% of American homes had the device, nearly
two and a half times the number of DVR households existing in January
of 2007.[4] For 95% of owners, the DVR was incorporated into their cable
or satellite set-top boxes, while only 5% owned a standalone DVR. Initial
direct sales of DVRs were limited because the technology did not offer new
programming; it only enhanced the consumptions of existing programming
sources (broadcast, cable, or satellite). The chief advantages of the product,
ease of time-shifting and commercial avoidance, were not enough to convince
masses of consumers to buy another set-top box with an additional monthly
fee. However, as TiVo and other DVR makers began to partner with the cable
and satellite providers, DVRs started to become part of cable/satellite package,
at first with a minimal additional fee, and later as part of enhanced services
used to attract new customers. For example, recently AT&T's video service
has advertised aggressively a DVR system that allows viewers to play recorded
programs in multiple locations throughout their homes. The message is that
viewers should switch to AT&T, not because it offers better programming or
a lower cost, but a better way to watch that programming: a superior DVR.

Ads for DirecTV and Dish Network routinely include offers of a free HD DVR along with free installation and discount pricing to lure customers from cable or their satellite rival. DVRs are increasingly a standard part of the cable package, just as remote control devices (RCDs) became part of the television set and VCRs in the 1980s. And just as RCD penetration moved from 29% to 84% in only 8 years,[5] DVRs are likely to begin a rapid ascent up the lazy S shaped diffusion curve.[6]

In March, 2009, Interpublic's Magna predicted that DVR penetration would rise to 44% of U.S. households in the next five years.[7] Considering the diffusion pattern of earlier television technologies and the rapid rise of DVR diffusion between 2007 and 2009, this prediction may be conservative. Looking at comparable five-year intervals shows that television penetration jumped from 34.2% to 83.2% between 1953 and 1958, VCR penetration went from 36% to 71.9% between 1986 and 1991,[8] and RCD penetration leaped from 29% to 77% between 1985 and 1990.[9] The average increase for these three technologies was 44.3%, more than three times the 13.5% predicted for DVRs. As with many innovations, now that DVR technology has become widely available and its benefits more apparent, smaller numbers of innovators and earlier adopters will give way to larger numbers of early and the late majority adopters. This diffusion pattern produces the lazy S shaped diffusion curve (with time on the horizontal axis and level of adoption on the vertical axis) common in diffusion of innovations studies.[10]

DVRs appear to be poised for rapid accent up the S curve. However, despite the general diffusion patterns of prior television technologies, the diffusion of DVRs is tied to how aggressively cable and satellite providers convert new and existing customers to DVR households. This "supply-side perspective" shifts the responsibility for adopting the innovation from the consumer to the television programming distributor, as the value of the device is evaluated by both the distributor and the customer.[11] As consumers switch television providers, they are likely to gain access to a DVR, just as television viewers acquired RCDs as a standard part of the new televisions they purchased in the 1980s and 1990s.

An aggressive cable/satellite approach is also aided by the use of nDVRs that require no in-home DVR, only interactivity with the cable/satellite provider. Programs are recorded by the cable/satellite operator for later playback by the viewer. The potential for nDVRs to rapidly increase DVR penetration has not gone unnoticed. In 2006, a consortium of networks and studios (ABC, NBC, CBS, Fox, Universal, Paramount, Turner and Disney) filed a copyright infringement suit to prevent Cablevision from using its remote server-based DVR. On August 4, 2008, the Second U.S. Circuit Court of Appeals ruled the use of nDVRs did not violate copyright law, and in June 2009 the U.S.

Supreme Court refused to consider an appeal. After the Court of Appeals ruling, Bernstein Research analyst Craig Moffett predicted that "in short order, effective DVR penetration could now jump to north of 60 percent."[12]

The Empowering DVR

In the digital video recorder, media researchers have found another new technological phenomenon to study. The DVR is a technology that, once again, has the potential to permanently alter how television is viewed, the programming and advertising strategies used by the U.S. television industry, and eventually the structure of that industry. At the current stage of adoption, the DVR's potential is the subject of much speculation, but research has focused primarily on commercial avoidance. Expectations about any changes brought to television by the DVR should reflect an understanding of the impact of earlier television technologies (RCDs, VCRs, multi-channel cable and satellite delivery systems). A combination of technological innovation and governmental deregulation over the last 30 years has destabilized the structure of the television industry, although a new oligopolistic structure has developed. The DVR, if widely adopted, may reshape the medium more completely than any previous "new" media technology in terms of media use, programming, and advertising.

The purpose of this chapter is to analyze the impact of the DVR on television advertising, programming practices, and viewers, while offering a few recommendations for industry practitioners. The audience is, of course, essential to any consideration of DVR effects on the industry. Both expected and unintended consequences are likely. The industry will respond in both predictable and unpredictable ways.

In discussing the importance of the remote control device,[13] it was argued that this seemingly innocuous add-on made possible the wide use of multi-channel television, ushering in a new way of looking at the television audience. The television viewer became the television user, constructing an individualized television mix by zipping, zapping, and grazing among the available options. There was also speculation about the prospects of a third generation, when interactivity and the Internet would combine, and when "the industry will have to conform to the desires of the television user."[14]

However, the DVR is more than a substitute for the VCR. It has the ability to both *find* (navigation function) content from a plethora of programming and *record* it easily (conduit function). In addition, the DVR is superior to both the RCD-equipped television and the VCR as a tool for the selective avoidance of commercials. Television RCDs allow viewers to switch to other channels during commercials (zapping), but the viewer risks missing some programming. Like VCRs, DVRs allow viewers to speed through (zipping) commercials, but much more rapidly than VCRs. In fact, many DVRs have a 30-second forward/reverse button that allows for the skipping of most advertising and promotional messages in recorded programs with a few button pushes. Users can easily reverse the action, if they accidentally skip some of the program. Viewers can also avoid commercials in an evening's worth of unrecorded programming by simply turning on their television and waiting 30 minutes or so to begin viewing. The DVR automatically records the programming when the set is turned on. The viewer can then rewind back over the first 30 minutes and start viewing. When a commercial break occurs they can fast-forward through the commercials. One early study reported that 88% of the commercials are skipped by DVR users,[15] while more recent research from TiVo's second-by-second Power-Watch puts the figure at 66% of all prime-time broadcast TV ads.[16] Even skeptical television executives believe 60% of DVR users zap ads.[17] This commercial-avoidance feature has been a focus of some product advertising for DVRs.

If the RCD combined with cable and VCRs brought television into a second generation, is the digital video recorder the "killer app[lication]" that will usher in the third generation? Will this technological genie, once out of the bottle, forever change television? At this time, nothing is certain, except that there is widespread speculation within the television industry that the DVR is contributing to an ongoing restructuring of the television industry. Tracey Scheppach, video innovation director at Starcom, predicted that "we'll see an acceleration of DVR-type technology, and that will destroy the present ad model."[18]

Nonetheless, attempts to analyze the effects of an emerging technology are problematic. Media analysts tend to overestimate the impact of new media technologies. We need to keep in mind the basic tenants of the diffusion of innovations. New products and services tend to diffuse at a slower rate than expected. The rate of diffusion increases if they offer relative advantages that are observable, are compatible with user needs and values, can be tried out before adoption, and are not too complex. In addition, cost is a key issue. Color television, VCRs, and remote control devices all diffused rapidly only after prices dropped significantly. Finally, existing industries will adapt to new circumstances, often blunting the diffusion of that new technology.

Winston's "law of the suppression of radical technologies" predicts that the existing powers in an industry, aided and abetted by the state, will be able to control the level of diffusion and use of a new technology.[19] This will help protect their vested interests. A corollary to this "law" is that if industry powers fail to suppress the radical potential of technology, they will simply buy it. For example, one of the major reactions to the growth of multi-channel cable was the lessening of barriers within and between broadcast and cable ownership to allow for more in-market and national integration of program conduits and programming ownership. If the old Big 3 television oligopoly (ABC, CBS, NBC) was to be no more, a new Big 5 oligopoly (Disney, Viacom, Time Warner, News Corporation, Comcast) would thrive by owning much of the cable and satellite competition. Similarly, the major developers of DVRs have corporate partners that are the major powers in the media entertainment industry.

Not surprisingly, the implications of DVRs are of increasing concern to industry analysts who realize that with the ability to easily find and store material from the increasing number of outlets, the television user can be more active and selective than ever. The active DVR user changes the traditional relationship between viewer and advertiser, as well as many established programming and program promotion strategies.

Advertising in the DVR Era

Because of its commercial significance, it is not surprising that DVR-enabled commercial avoidance has received the greatest attention from both academic and industry researchers. In reviewing this emerging research, several conclusions are evident. First, one of the two most important motivations for owning a DVR is the ease of commercial avoidance; the other is ease of time-shifting, recording programming for viewing at a more convenient time.[20] According to a 2005 online survey by the media agency MindShare, 79% of responding owners said they purchased a DVR because it helps them skip commercials.[21] Second, viewers of DVR recordings skip ads at a substantial rate. A recent study of second by second TiVo data found that 66% of all recorded ads were skipped,[22] while figures compiled from November 2007 and 2008 data in Nielsen's recent report on DVR households in the U.S. show an average of 56.4% of commercials were not viewed.[23] Third, all non-program content is not equally affected. Commercials at the start and the end of the commercial pod are more likely to be seen. This is one reason why advertisers are already paying more for ads at the start and end of commercial pods. In addition, network promos were skipped less than other ads.[24] Fourth, ad skipping is not the same for all networks, because DVR time-shifting varies considerably by network. Programs on broadcast networks were more frequently recorded than on specialized cable networks. For example, a three-year study comparing DVR to non-DVR households found DVR users time-shifted 42% of CBS programming versus 18% of Lifetime's and 10% of the Food Network's programming.[25] Finally, the commercial television industry, especially those entities that are still primarily supported by spot advertising, are deeply worried about the long-term threat posed by DVRs.[26] One study comparing DVR and non-DVR households found the purchase of "Pacesetter" package-goods brands to be 5% lower in DVR households with 20% of all brands showing a loss in sales volume. However, there were no losses for products that relied less on spot advertising and had more diversified media plans.[27] In short, DVRs hurt brands that rely too heavily on traditional spot advertising.

The Impact of DVRs on the Television Commercial

Using Nielsen prime-time data from November 2007 and 2008 for ABC, NBC, CBS, and Fox, we can calculate the amount of current loss in attention to advertising and project that to a DVR-ubiquitous future.[28] DVR owners watch an average of 9.9 hours per week of recorded programming, representing 29.8% of their total viewing. During this viewing, they skip an average of 56.4% of the commercials. Thus, for this group, commercial viewing is reduced by 16.8% during their total television viewing. If this figure is multiplied by the current DVR penetration rate (30.6%), the best estimate of the current impact of DVRs on commercial viewing in network prime-time programming is a reduction of 5.1%. If the commercial skipping and time-shifting patterns documented by Nielsen continue, the percentage of commercials avoided due to DVR use would reach 16.8% if DVR penetration reaches 100%. This figure is only a projection and is based on the assumption that the total amount of time-shifting and commercial avoidance reported by Nielsen for these two months is an accurate estimate of future DVR behavior. It can be concluded that DVRs currently are having a significant (about 5% reduction) effect on the value of commercials in network prime-time programming. As DVR penetration increases, the value of spot advertising will continue to decline, but spot ads will still be unaffected by DVRs in over 83% of network prime-time programming. By themselves, DVRs will be a significant blow to, but not the end of, television commercials.

Also, limiting the potential negative impact of DVR-inspired advertising-avoidance are research findings that show advertisers may receive some benefit from zipped (fast-forwarded) ads. Specifically, visual content in the middle of the screen during the fast-forwarding process is perceived by viewers. Significantly, viewers appear to pay closer attention to the screen during zipping than during normal viewing, and this can have a positive impact for the advertiser.[29] However, the audio content is lost and the visual message greatly simplified, promoting mostly brand awareness. Marc Goldstein, head of MindShare North America (a major media buying consultant), reports a "30-second Pontiac commercial [he zipped] was intended to introduce the new brand, show that it has two doors and a 250-horsepower engine and gets 35 miles to the gallon," but all he learned was "there was a Pontiac commercial that I fast-forwarded."[30] Also, limiting the potential impact of zipped commercials is the use of the 30-second button available on some DVR remotes. This button skips content in 30-second units, meaning that virtually all of the content in a 4-minute commercial pod can be eliminated with eight pushes of this button. Commercials are not "zipped" but "zapped."

In the mid-2000s, the debate over the value of the commercial content in DVR programming triggered confrontations between networks and advertising agencies over how Nielsen would count DVR-viewed programming in its ratings. Advertising agencies argued that the extent of ad zipping made DVR viewing worthless, while networks insisted there was still value. A compromise was negotiated and currently Nielsen includes DVR viewing in its ratings if it occurs within three days of the telecast and "accounts for fast-forwarding as well as channel switching."[31]

Responses to Commercial Avoidance

Despite the enhanced threat from the DVR, the television industry has been weathering threatening storms for the last 20 years. In the first generation of television characterized by limited viewer choice, simplistic audience measurement tools, and the Big 3 oligopoly, advertisers were confident that their commercials reached most of the people watching the program in which the ads appeared.

In the second generation of television, several factors have shaken advertiser confidence. Audiences have access to vastly increased program choice. More precise audience measurement tools have become available. Most important, a wide range of tools that enable the audience member to navigate the video landscape are now available. Thus, most advertisers have had to revise and rethink the value of television commercials.

These factors and the diffusion of VCRs, cable, and RCDs led the advertising industry to take several interrelated actions:[32]

1. Spreading ads to the new video outlets.
2. Emphasizing audience demographics and psychographics rather than mainly audience size in media buying decisions.
3. Enhancing the production value of spots with more contemporary music and humor, fast editing, and offbeat narrative, often in shorter

spots, to quickly grab the attention of the restless viewer. For example, one element of conventional wisdom in television promotion is that the viewer must be grabbed in the first 3 seconds or s/he is likely to change the channel.[33]

4. Increasing the emphasis on the brand qualities of the product or service being advertised. This emphasis is used in integrated marketing communication (IMC) schemes that use multiple platforms of advertising combined with traditional marketing plans and public relations campaigns to get the message to the intended target. Branding is a large part of IMC because it is assumed that brand identity and image are ingrained in the mind of the consumer and can be activated with limited exposure to an associative device such as a logo.[34]

5. Making more advertising "zap-proof" by integrating it into the programming, so it can only be avoided if parts of the program are avoided. One of the explanations for the increase in sports programming is that this programming provides multiple opportunities for zap-proof ads. These opportunities include the sponsorship billboards keyed into the telecast at the beginning of the event, venue signage, on-air sponsorship of specific game elements, and virtual signage that can be inserted in the picture. Of course, sports telecasts have always incorporated some of these strategies but they have become increasingly prevalent. For example, McAllister's review of the college football's 2007 national championship game found 80% of the telecast included some kind of on-screen advertising.[35]

As DVRs and other forms of new media technology move television into its third generation, each of these second-generation strategies will take on increasing importance, as the value of traditional spot advertising shrinks. In addition, as will be discussed later in this chapter, television networks and stations will more aggressively pursue additional revenue streams to compensate for lost spot-advertising revenues.

Programming in the DVR Era

DVRs present a new challenge to programmers. Since the viewer now records programs by title from genre lists or through key word searches, selection has little to do with the appeal of adjacent programs. Over a typical evening, viewers watch programs that may have aired at any time over the previous few days or earlier that evening. The programs are watched in whatever order the viewer wants. There is no hammock, tent-pole, or lead-in program effect.

In addition, viewers are not limited to selecting the best program alternative available at a particular time. They select from a list of programs aired over several days. Instead of choosing the *least objectionable program*, they are selecting from among their *most desired programs*. In short, for DVR-recorded programs, scheduling is unrelated to selection.

As the producers of the audience for sale to advertisers, programmers are more vital to the success of the industry than ever. As the explicit partners of the advertisers, programmers in the DVR era can and will cooperate with advertisers by allowing more sponsor identification, product placement, virtual signage and branding.

Programming outlets do this by stressing their own brand strengths and packaging. While the measurement of brand equity in television is highly problematic,[36] there is little doubt that major television network outlets are leveraging their brand identities. Networks have connected their brands to other television outlets (MSNBC, ESPN2, ESPNEWS), other media (*ESPN-The Magazine*, CBS Sportsline), other businesses (ESPN Sports bar/restaurants), and their own on-air promotional campaigns. These branding strategies make financial sense on two levels. First, leveraging a strong brand can bring in revenue from new sources even if the primary branded entity (the network) is lagging. Second, branding can help convince viewers that the branded entity is a safe harbor in a time of confusion, a safe place to find television entertainment or information that is desirable because it comes from a trusted brand.

Another element of brand strategy is consistency of schedule and specific programs. This helps create appointment television: programs so compelling that viewers will make time in their schedules to view them. Networks increasingly are highlighting their most popular or critically acclaimed programs for special emphasis. Of course, appointment viewing/usage has always been the basis of all out-of-home media and entertainment (movies, concerts, theatre, amusement parks).

Despite the new emphasis on brand extension and promotion, the major television providers have not abandoned established scheduling strategies that seek to control flow. There is simply too much evidence that such strategies as hammocking, tent-poling, and lead-in continue to work even in a time of declining ratings for most programming. In fact, declining ratings are a key reason for maintaining such practices. Now that every rating point or even fraction of a point in every time period is important, programmers must schedule carefully to keep viewers tuned to their networks through a series of programs. However, as DVRs become more common and less programming is watched at the time it was originally aired, these flow-control strategies will become less important.

Much of the value of live programming such as sports, news, reality shows and award programs comes from the audience's experience of a real-time event. Live events are by definition appointment television. DVR users are much less likely to record live programming and thus much more likely to flowthrough to programs and advertising that follow live events. There is little doubt that live programming will have increasing value as DVR penetration rises.

DVR Viewing

Hard research on the impact of DVRs on viewing patterns is only starting to emerge. In a recently published observational and interview study using a small convenience sample, Smith and Krugman found that participants were highly attentive when watching DVR-recorded programming, more attentive than when watching live television or even VCR rentals.[37] However, advertising avoidance among DVR users was even more pervasive, with participants either fast-forwarding, switching channels or leaving the room during 95% of the advertising segments. DVR users watched the screen 94% of the time they were not engaged in other competing activities (e.g., reading, interacting with pets, checking cell phone messages, etc.), but still had their eyes on the screen 72% of the time they participated in competing activities. In-depth interviews revealed that "DVR owners develop a sense of 'über-control' over television" and "a clear synergy takes place when consumers combine selection, scheduling, sorting, pacing, pausing and rewinding of programming," and avoiding commercials. The "viewer transitions from reactive content receiver to proactive media decision maker."[38] Although limited by a small convenience sample, these findings are consistent with earlier research that found DVR owners got more enjoyment and greater control out of television.[39]

Nielsen's recent DVR report on prime-time network viewing also shows how audiences use the technology.[40] Not surprisingly, DVR playbacks occurred when most viewers had time to watch TV, during weekday evenings and weekend days. The lowest DVR viewing levels occurred on Saturday evening, a traditional low viewing time. DVR playback peaked in the fall months when networks premiered new episodes. Programs recorded in the first hour of prime time were more likely to be played the evening of their recording and more likely to be played at all than programs in the last hour of prime time. DVR use showed some demographic differences, with the heaviest time-shifters being women and viewers over 50. These are groups that generally watch more network prime-time television as well. Males 35-49 were the least likely to be heavy time-shifters. Viewers were able to use their DVRs to watch more episodes of their favorite series, seeing more original telecasts of their preferred programs. Nielsen also reported in June 2009 that the most frequently re-corded broadcast network programs (DVR viewing as a percent of total view-ing) were virtually all episodic dramas and comedies.[41]

As noted before, a study by Information Resources Inc. comparing DVR vs. no-DVR households found more time-shifting of network programs (42% for CBS) than specialized cable network programs (10% for the Food Network and 18% for Lifetime). Network promos seemed more popular than ads, with 60% of the promos in the first position of a commercial pod viewed versus 45% of commercials in the same position.[42] Finally, several studies found that DVR households watch more television that non-DVR households.[43]

Program and Channel Promotion

Although network promos are more popular with DVR users than commer-cials, they are still avoided in substantial numbers by DVR users. Program-mers have long relied on promos placed in commercial breaks to introduce new network programs and promote new episodes of established shows. If commercials imbedded in DVR-recorded programs are zipped as frequently as early research suggests, promos will be viewed less frequently. Thus, DVRs may reduce the effectiveness of one of the most important program audience-building tools. As in the case of advertising messages, however, zap-proof methods of in-program promotion, such as animated promos for upcoming programs on screen while a program airs, will continue to increase in impor-tance. In addition, promotion increasingly will use a variety of "platforms" to reinforce both brand identity and specific messages (i.e., ESPN's sports bars and magazine, Disney's various channels, theme parks, etc.).

The potential of DVRs to change television has led many media corpora-tions to invest in the technology. "If you can't beat 'em, buy 'em." At the dawn of DVRs, Disney and NBC bought equity interest in both ReplayTV and TiVo, while AOL TW was the lead investor in Moxi.[44] Discovery Networks, Blockbuster, and British Sky Broadcasting (BSkyB) also had investments in the DVR industry. Unlike most major film studios, which initially ignored television, media companies hedged their bets with investments in DVR technology.

DVRs: Audience Implications

Will DVRs be the end of grazing? One of the most important consequences of the diffusion of RCDs over the last two decades has been the increase in channel changing by more active viewers. Academic and industry studies have reported between 5% and 20% of viewers are switching among several channels instead of viewing only one channel at a time, while recent industry figures show 6% to 8% of viewers use their RCDs to zap commercials.[45] Researchers have identified two patterns of frequent channel changing: grazing or channel surfing where viewers move among many channels, watching each for a short time, and multiple program viewing, where viewers focus on the content of two or three channels (e.g., watching two baseball games at once).

This rapid shifting from one channel to another is similar to the magazine and newspaper flipping discussed by Stephenson in his elaboration of the play theory of mass communication.[46] Indeed, one study found a correlation between consumers' tendency to "flip" through content of several different types of print and electronic media.[47] Stephenson argues that this kind of media use provides communication pleasure of its own, independent of the content consumed. The RCD made both grazing and multiple program viewing easier, and cable/satellite television gave these venturesome viewers more programs to sample.

Recent research shows higher levels of attention to DVR recordings than to conventional television because the programming is perceived to be of greater value.[48] Grazing behaviors have often been attributed to viewers who failed to find a particular program that interested them, but continued to watch television, gaining gratification for many short program segments. DVRs allow viewers to record hours of viewer-selected programming that is available on demand. The viewing boredom and program dissatisfaction that stimulated much of the RCD grazing of the past should be reduced by the use of DVRs. If so, some of the loss in commercial exposure due to DVR zipping many be replaced by less commercial zapping from RCDs.

Selective Exposure and Avoidance

Several studies of RCD gratifications have found that the selective avoidance of unpleasant stimuli, including politicians, political ads, and news reporters is an RCD motivation of moderate strength and is significantly related to RCD use.[49] In studies of undergraduates and adults, researchers found between 40% and 60% of the respondents agreed with statements supporting selective avoidance.[50] These studies suggest that many viewers use the remote control tool to reject unwanted stimuli.

While the RCD facilitates selective *avoidance*, the digital video recorder is, above all else, a tool for selective *exposure*. The recording device combined with the software controls allows viewers to select and easily record only the programming that interests them. The "season pass" feature allows viewers to select certain programs for automatic recording each time the program is aired for an unlimited time period. DVR software also offers a variety of search options to find desired programming, including a "wish list" option that allows viewers to find programs featuring particular performers or directors. Indeed, watching a DVR means watching only the programs previously recorded and listed on its menu. Thus, the digital video recorder's major advantage for viewers, in addition to commercial avoidance, is its selectivity power: viewers may select only their most desired programming from a vast and often confusing array of cable and satellite options with little exposure to any unwanted options.[51]

As digital video recorders become commonplace, then television viewing will become more personalized as viewers typically encounter only programming that they have selected in advance and that is presented to them as options on their television menu. Clearly, major news events will receive wide coverage from multiple television channels and thus will be widely experienced despite the presence of DVRs. But less dramatic and compelling stories, personalities, and ideas will be seen by only those who choose them.

As was noted about RCD-induced selective avoidance, increased selectivity means more personalized television use and less sharing of common information, opinions, and concerns within the general population. The self-serving DVR user may be less likely to listen to or even be aware of the arguments of groups that oppose his or her point of view, in effect withdrawing from democratic discourse. The constant selection of similar one-sided viewpoints is likely to harden attitudes and reduce tolerance for other positions. Even more unsettling, programs focusing on the uncomfortable realities of modern life, such as crime, poverty, and social injustice, can be ignored because they are never recorded. In an era where most of the population receives their news from television, social problems become increasingly hidden problems. Thus, although it may be personally desirable, an increased capacity for selective television programming represents a pressing problem for the television industry and perhaps for the larger society.

What to Do about DVRs?

Television networks and stations first must avoid overstating the extent of the DVR's impact. The DVR has empowered viewers to be more selective in their television viewing and has added a new tool in their resistance to the commercial. However, an analysis, based on the best Nielsen data in the public record, suggests that DVRs are only reducing the amount of commercial exposure in network prime-time programs by about 5% and that figure would only reach about 17% if all households had DVRs and used them to skip to the same extent that DVR households currently skip them. Clearly, these figures represent a substantial impact, but not the end of commercials as an effective advertising tool or a form of financial support for television.

A crucial television industry strategy to weaken the DVR threat is simple to state but difficult to execute: provide appointment programming. DVR ad avoidance is most intense in programming that can be readily time-shifted because nothing in the programming requires the viewer's immediate attention. Standard recorded entertainment programming (dramas, situation comedies, movies) that offer no anticipated resolution can be easily recorded on a DVR for consumption at a more convenient time. As noted earlier, sports, news and some reality programming provide time sensitive information that discourages delayed viewing. In addition, the more specialized programming offered by targeted cable channels (Food Network, Lifetime) appear to be less frequently time-shifted than broadcast networks. A heightened interest in a particular program leads to less time-shifting and greater attention to commercials.

Although all research on DVR use is preliminary at this juncture, it appears that programs in the last hour of prime time are most likely to be recorded but not viewed, while those recorded in the first hour of prime time are more likely to be viewed and viewed that same evening. This suggests that programs with the most immediate value or with strong interest (appointment TV), programs most likely to be watched live, should be scheduled in the last hour of prime time to reduce the impact of lost viewing due the failure to play DVR recordings.[52] Networks will also need to adjust expectations for ratings generated from repeats of original episodes, since more DVR viewers will have seen the original airings of their favorite shows.

For local stations, appointment programming has meant local news, which has expanded substantially in the second generation of television. Because of its immediacy, news is not as likely to be time-shifted as entertainment programming. However, because of the extensive time already being devoted to news at many stations, it is not likely to increase much more. Other sources of live local programming, including local sports, talk shows or local versions of popular reality shows, also could be DVR-resistant. Local programming would provide numerous opportunities for sponsorship and local product placement, as well as promotion for the originating station. The availability of lower cost production facilities and the acceptance of new media technologies such as "YouTube" video aesthetics may help reduce costs and make expanded local programming more possible than in the "broadcast standard" era.

Networks and stations will need to experiment with commercial placement within pods to help reduce DVR commercial zipping and zapping. Placing the most desirable non-program material (program promos, motion picture ads) at the start of the pod could lure some would-be DVR zippers into watching most, if not all, of the commercial pod. In addition, the introduction of even more entertaining/involving commercials would help keep viewers' fingers off the remote buttons. Finally, media companies need to work with DVR manufacturers to reduce the number of DVRs with 30-second skip buttons, since this makes commercial zapping much easier and eliminates virtually all of the benefits of fast-forward mode viewing.[53] However, techniques for fighting DVR commercial avoidance are not enough. The television industry must chase additional sources of revenue beyond spot advertising.

Both networks and other broadcast station owners are more aggressively pursuing retransmission consent fees from cable MSOs and satellite service providers that currently pay ESPN over $4 for each subscriber.[54] Recently, Fox demanded from Time Warner cable more than a dollar per subscriber for retransmission of its local station signals, although it likely settled for substantially less.[55] The era of the dual revenue stream for cable networks and, the single (advertising-only) stream for broadcast entities is over. In 2009, stations received $739 million for retransmission rights, constituting 4.2% of their revenue, and that figure is expected to rise to 8-9% of revenues by 2015.[56] Consumers viewing broadcast networks and their local stations on cable or satellite will likely pay more for the privilege of watching their once "free" (advertiser-supported) television on cable/satellite services.

In addition, both networks and local stations will need to continue drawing more revenues for their programming from non-spot advertising sources such as downloads on their own website or other video websites (Hulu, Fancast, Veoh, etc.), selling through Web retailers (iTunes, Amazon.com), DVD sales, and mobile video downloads.

DVRs are the next step in a 30-year evolution that has given television viewers more power to choose what they want to watch and when they want to watch it. Following the diffusion of multi-channel cable television, VCRs, RCDs, and satellites, the DVR is both an outlet for programming and a way for the television user to navigate and organize an increasing number of program choices.

Although currently DVRs are in less than a third of U.S. households, they are poised for a rapid expansion as cable and satellite services provide a supply-side push.[57] The DVR is increasingly regarded as a "gateway" technology, the type that will lead users to new services such as video on demand. As such, as they become a standard feature within the set-top boxes that enable consumers access to digital cable and satellite. As an add-on perceived as entailing no extra cost to a new television, RCDs diffused rapidly in the 1980s. Analysis of the diffusion of earlier television technologies (TVs, VCRs, RCDs) saw rapid diffusion after the 30% figure that DVRs now have. In addition, as an important navigation device, DVRs are tools designed for "lazy interactivity:" popular devices that promote quick decisions, cater to short attention spans, and supply instant gratification.[58]

The DVR is part of an ongoing trend rather than an isolated breakthrough. Much more study needs to focus on how people are making use of the various navigation and program expansion technologies in their homes and how these technologies may affect their media use and interpersonal relationships. Neuendorf's call for "reopen[ing] dialogue on the basic question of what constitutes an audience for contemporary media products" seems particularly important as technological diffusion alters the traditional parameters of audience/outlet relationships.[59] Conversely, sports, news programming, and reality programs/contests offer programs that are best enjoyed in real time. Audiences want to watch the Super Bowl as it happens, but they can watch the next episode of *Two and a Half Men* at their leisure. As DVRs become as ubiquitous as RCDs and the time-shifting of prerecorded programming becomes commonplace, broadcasters will need to draw revenue from non-spot advertising sources to support all of their programming. Indeed, this process is well underway as the second generation of television has moved into the third. Networks now exploit their programs through Internet streaming that forces viewers to watch commercials before the programs begin, and also through a variety of aftermarket sales: U.S. and foreign syndication, DVD sales, mobile video, and the Internet.

As for the industry, history has shown that new viewer empowerment will be forcefully countered. The industry already has responded by reconstituting its oligopoly economic structure, increasing its ownership of newer technologies (including DVRs), and promoting more integration between programming and advertising.[60] Ironically, because of DVR-enhanced commercial-avoidance, viewers will face either more intrusive, zap-proof advertising or an ever increasing bill for the programs they watch.

Notes

1. Robert V. Bellamy, Jr., and James R. Walker, *Television and the Remote Control: Grazing on a Vast Wasteland* (New York: Guilford, 1996).

2. Anthony Bianco, "The Vanishing Mass Market," *BusinessWeek Online,* http://www.businessweek.com/magazine/content/04_28/b3891001_mz001.htm (12 June 2004).

3. M. Bjorn von Rimscha, "How the DVR Is Changing the TV Industry—A Supply-Side Perspective," *International Journal on Media Management* 8, no. 3 (2006).

4. "How DVRs Are Changing the Television Landscape," Nielsen, blog.nielsen.com/nielsenwire/wp-content/uploads/2009/04/dvr_tvlandscape_043009.pdf (April 2009).

5. Bruce C. Klopfenstein, "From Gadget to Necessity: The Diffusion of Remote Control Technology," in *The Remote Control in the New Age of Television*, eds. James R. Walker and Robert V. Bellamy, Jr. (Westport, CT: Praeger, 1993), 23–39.

6. Everett M. Rogers, *Diffusion of Innovations*, 4th ed. (New York: Free Press, 1995).

7. Brian Steinberg, "Ad Skippers Beware: Ask.com Going After You with TV Crawl," *Advertising Age*, 2 March 2009, 4 and 33.

8. Christopher H. Sterling and John Michael Kittross, *Stay Tuned: A History of American Broadcasting*, 3rd. ed. (Mahwah, NJ: Erlbaum, 2002), 864 and 866.

9. Klopfenstein, 1993, 33.

10. Rogers, 1995.

11. von Rimscha, 2006.

12. Anthony Crupi, "Remote Control," *MediaWeek*, 11 August 2008, 4.

13. James R. Walker and Robert V. Bellamy, Jr., eds. *The Remote Control in the New Age of Television* (Westport, CT: Praeger, 1993); Bellamy and Walker, 1996.

14. Bellamy and Walker, 1996, 146.

15. M. Lewis, "Boom Box," *New York Times Magazine*, 13 August 2000, 36-41, 51, 65-67.

16. Crupi, 2008.

17. Steinberg 2009.

18. Crupi, 2008.

19. Brian Winston, *Misunderstanding Media* (Cambridge, MA: Harvard University Press, 1986).

20. John Consoli, "MindShare: Consumers Buy DVRs to Skip Ads," *MediaWeek*, 19 December 2005, 6–8; Douglas A. Ferguson and Elizabeth M Perse, "Audience Satisfaction Among TiVo and ReplayTV Users," *Journal of Interactive Advertising* 4, no. 2 (Spring 2004): 1–8.

21. Consoli, 2005.

22. Crupi, 2008.

23. "How DVRs Are Changing the Television Landscape."

24. Jack Neff, "Study finds mixed DVR effects," *Advertising Age*, 24 March 2008, 8.

25. Neff, 2008.

26. John Consoli and Anthony Crupi, "DVRs Make Their Presence Felt," *Advertising Age*, 30 April 2007, 8–10; Neff, 2008; Steinberg, 2009; Kenneth C. Wilbur, "How the Digital Video Recorder (DVR) Changes Traditional Television Advertising," *Journal of Advertising* 37, no. 1 (Spring, 2008): 143–49.

27. Neff, 2008.

28. "How DVRs Are Changing the Television Landscape."

29. S. Adam Brasel and James Gips, "Breaking Through Fast-Forwarding: Brand Information and Visual Attention," *Journal of Marketing* 72 (November 2008): 31–48; Caleb Siefert, Janet Gallent, Devra Jacobs, Brian Levine, Horst Stipp, and Carl Marci,

"Biometric and Eye-Tracking Insights into the Efficiency of Information Processing of Television Advertising during Fast-Forward Viewing," *International Journal of Advertising* 27, no. 3 (2008): 425–26.

30. "The Upfront and the Unknown," *Broadcasting & Cable*, 8 May 2006, 25.

31. Marc Berman, "Live, From . . . ," *Media Week*, 29 October 2007, 38.

32. Bellamy and Walker, 1996.

33. Robert V. Bellamy, Jr. and J.B. Chabin, "Global Television Promotion and Marketing," in *Promotion and Marketing for Broadcasting and Cable, 3rd ed.*, eds. Susan T. Eastman, Douglas A. Ferguson and Robert A. Klein (Boston: Focal Press, 1999), 211–32.

34. Robert V. Bellamy, Jr., and Paul J. Traudt, Television Promotion as Branding. In *Research in Media Promotion*, ed. Susan T. Eastman (Mahwah, NJ: Erlbaum, 2000), 127–59.

35. Matthew P. McAllister, "College Bowl Sponsorship and the Increased Commercialization of Amateur Sports," *American Behavioral Scientist* (2010): forthcoming.

36. Bellamy and Traudt, 2000.

37. Sarah M. Smith and Dean M. Krugman, "Viewer as Media Decision-Maker: Digital Video Recorders and Household Media Consumption," *International Journal of Advertising* 28, no. 2 (2009): 231–55.

38. Smith and Krugman, 2009, pp. 249–50.

39. Ferguson and Perse, 2004.

40. "How DVRs Are Changing the Television Landscape."

41. Bill Gorman, "*Dollhouse* Had Largest Share of Viewing From DVRs, Did That Save The Show?" *TV by the Numbers*, tvbythenumbers.com/2009/06/18/dollhouse-had-largest-share-of-viewing-from-dvrs-did-that-save-the-show/21052 (18 June 2009).

42. Neff, 2008.

43. Neff, 2008; "How DVRs Are Changing the Television Landscape."

44. P. J. Brown, "PVRs: Content Control," *Cahners/TVinsite*, www.tvinsite.com/index.asp?layout=print_page& doc_id=&articleID=CA19042 (10 July 2002).

45. Bellamy and Walker, 1996; Bill Carter, "DVR, Once TV's Mortal Foe, Helps Ratings," *New York Times*, 2 November 2009, B1.

46. William Stephenson, *The Play Theory of Mass Communication* (Chicago: University of Chicago Press, 1967).

47. Bellamy and Walker, 1996.

48. Ferguson and Perse, 2004; Smith and Krugman, 2009.

49. James R. Walker and Robert V. Bellamy, Jr., "The Gratifications of Grazing: An Exploratory Study of Remote Control Use," *Journalism Quarterly* 68, no. 3, (1991): 422–31; James R. Walker, Robert V. Bellamy, Jr. and Paul J. Traudt, "Gratifications Derived from Remote Control Devices: A Survey of Adult RCD Use," in *The Remote Control in the New Age of Television*, eds. James R. Walker and Robert V. Bellamy, Jr. (Westport, CT: Praeger, 1993), 103–12; Lawrence A. Wenner and Maryann O'Reilly Dennehy, "Is the Remote Control Device a Toy or Tool? Exploring the Need for Activation, Desire for Control, and Technological Affinity in the Dynamic of RCD Use,"

in *The Remote Control in the New Age of Television,* eds. James R. Walker and Robert V. Bellamy, Jr. (Westport, CT: Praeger, 1993), 113–34.

50. Walker and Bellamy, 1991; Walker, et al., 1993.

51. Smith and Krugman, 2009; Ferguson and Perse, 2004.

52. "How DVRs Are Changing the Television Landscape."

53. Brasel and Gips, 2008; Siefert, et al., 2008.

54. Melissa Grego, "Retans: The Bloody Battle to Save Broadcast Television," *Broadcasting & Cable,* 14 December 2009, 10–13.

55. Brian Stelter, "Time Warner and Fox Reach a Cable Deal," *New York Times,* 2 January 2009, B1–2; Grego, 2009.

56. Grego, 2009.

57. Von Rimscha, 2006.

58. C. Waltner, "Many Challenge Future of Interactive TV," *Cahners/TVinsite,* www.tvinsite.com/index.asp?layout=print_page& doc_id=&articleID=CA33947 (7 December 1998).

59. Kimberly A. Neuendorf, "Viewing Alone? Recent Considerations of Media Audience Studies," *Journal of Broadcasting & Electronic Media* 45, no. 2 (Spring 2001): 352.

60. McAllister, 2010.

4

The Obstinate Audience Revisited

The Decline of Network Advertising

Douglas A. Ferguson

A DVERTISING IS EFFECTIVE WHEN ITS MESSAGE cannot be avoided. In an analog world, advertising avoidance is of little consequence, but digital technologies are making inroads. Until the dawn of the twenty-first century, media have thrived using models supported by subscriptions and advertising, sometimes just the latter. Advertising-supported mass media can no longer thrive or perhaps survive in a digital world where (1) audiences are empowered to avoid advertising easily and (2) where the number of choices greatly exceeds the amount of discretionary time spent with media. Borrowing from Bauer's transactional model of audience influence, consumers of broadcast messages are exerting their collective power to renegotiate the playing field.[1] The audience is now far more powerful than the broadcasters, in part because they have become active participants in how media content is chosen, distributed, and (in the case of user-generated content) produced.

Compounding the problem is that audiences (especially the young) are increasingly expecting most content for free, with or without advertising. Furthermore, the adopters of new media have not only learned *how* to expect their media, but also *when* and *where*, that is, all the time and everywhere. As a result, a fundamental shift has taken place.[2] For broadcast media, the old business model is badly broken, and for print media, the medium itself is seemingly lost. The root causes are an ever-expanding cornucopia of choices, abetted by convenient access to free content on the Internet.

The scope of this chapter is the United States. Global differences are important, but not in these pages. The reader is referred to other sources, available by searching "global advertising outlook" on the Web. One good reference is a 2007 IBM monograph cleverly entitled "the end of advertising as we know it."[3] European countries were less reliant on advertising until twenty years ago, with many media state-controlled. For those nations, the disconnect may be less intense.

Network advertising is not healthy. For example, the average unit cost continues to decline as the audience size shrinks at a faster pace.[4] Several important changes in the media landscape comprise the shift away from advertising-based media. Certainly the appearance of additional revenue streams influenced network (and local) broadcasters, as the audience slowly learned to embrace pay-per-view and other forms of premium television provided by multichannel sources like cable and satellite. But what is not so obvious is that the gradual shift of audience preferences, coupled with digital media and more choices, created a loss of urgency and the expectation that media, like Web content, should be free. Moreover, media content should be free of advertising, free of market barriers to entry, and free of limited choice.

The Loss of Urgency

New media have served to disconnect the urgency of consuming media from the economic goals of media producers. Media consumption depends on the perception of content availability, or *media urgency*. Content was never king because of its inherent qualities but instead because of its relative scarcity. The concept of media urgency is relatively unstudied as a variable in media use, but it is easy to understand. Media urgency is the sense that if content is not consumed during some near-term window of opportunity, then future opportunities are constrained. Each medium has its own degree of urgency based on the duration of its window. Each window offers different economic benefits to the content provider: the earlier the consumer feels the urge to consume content, the higher the price (and the greater the reward to the provider).

For example, video content has multiple windows. A major (i.e., big-budget) motion picture has a theatrical window, followed by several in-home options. Before the advent of home video and cable television, feature films had few windows: release in large cinema venues, re-release in small movie theaters, eventual availability on network television, and final syndication to television stations to play late at night or on weekends. Television stations unaffiliated with one of the former-big-3 networks would play movies several times per day. After cable television reached a majority of homes, specialized channels provided another method for viewing those older films that people had missed during theatrical release, either because they felt little urgency or were prevented by other circumstances (e.g., birth age). Pay movie channels (e.g., HBO), video on demand (VOD), video rentals, and direct sell-through provided more options to Hollywood producers. Nowadays, people decide when they might see a movie ("I'll wait until it's on free TV") and producers decide whether to save the film for summer release or offer it up as direct-to-video fare. Each movie thus has a perceived urgency for any given consumer. Some people are motivated to see all the important movies right way and other consumers could not care less about cinema.

Urgency is influenced by the perceived need to discuss a media product with one's close associates, as well as a host of media gratifications for enjoying media content.[5] A new best-selling book from a popular author will cause excitement for the writer's fans, sometimes getting them to make a mental appointment to be at the bookstore on the day of release. Or other fans who are less rabid may wait until a less expensive copy appears in paperback or put their names on a waiting list at the public library. Urgency depends on the author or the book and the motivation to read it now rather than later.

Advertiser-supported media like television, radio, and newspapers have been much more perishable commodities than books until recent years, certainly after the turn of the century. If you missed an episode of *Bewitched* in 1966, you had to hope that it would be one of the 13 reruns that summer (or wait until it appeared in syndication years later, assuming a sufficient number of episodes). Today, if you miss an episode of a contemporary comedy half-hour, you can buy the whole season on DVD at a retail store (or online), even for some older classic sitcoms. It no longer matters that your favorite show has gained enough momentum to appear in broadcast or cable syndication. Nowadays, broadcast television syndication is mostly first-run programming.

The diffusion of the digital video recorder (DVR), estimated to be over 33 percent of homes in 2009,[6] allows fans of television shows to get a "season pass" so that every episode will be recorded without regard to urgency. So-called "must-see TV" is a concept locked in the 1990s, in the days before DVRs. Most viewers may not want to watch a recorded version of a live sports contest or an awards program (especially if they learn the final outcome before they get a chance to see it). Urgency is greater for live events, especially those that provide conversational currency the next day in the workplace.

Loyal listeners to radio hosts are sometimes motivated by urgency. If you missed hearing Rush Limbaugh on his radio show in 1993, then you were unlikely to get a second chance. Today, however, you could subscribe to his 24/7 service and have access to Internet audio archives of any show you missed. The sense of urgency for his legion of radio fans is reduced by technology, at a price. Other syndicated radio programs (e.g., *The Bob & Tom Show*) offer similar subscriptions to archival audio. Although the producer rescues the value, the medium itself does not share the revenue. The radio stations in this example simply lose some of their audience because listeners are less motivated to be part of the live audience.

Newspapers are more or less permanent, available at the library, but usually home-delivered copies are discarded or recycled. If you missed a fascinating series of local newspaper articles in the 1980s, the effort to retrieve another opportunity to read them may not have been worth your trouble, at the time. With the Internet, however, the chances of locating an old newspaper article are enhanced, even without cost if the request is made within days of missing that edition. Most newspapers now recognize the value of archival material older than a week or a month and charge a premium for adding another window of opportunity.

The Internet itself is the least urgent of all media, potentially affecting urgency of other media that migrate into the "new media" environment. The Web is prone to automatic self-archival (although some sites seek to monetize the ability to search archives). If urgency is an important motive for media gratification, then the Internet is disrupting media consumption. From the beginning of the Internet, the culture of the Web was to set information free, to detach content from cost. That culture has embedded itself in the media consumption mind-set of youthful consumers, for whom everything should be free, even at the risk of denying the copyright holders' intellectual property. Yesterday's army of teenage MP3 downloaders has become today's media consumers.

Perhaps urgency is related to one's internal metronome, or the speed with which one regulates daily activities. We suspect that regional differences account for speed of walking and driving. Northerners are typically in more of a hurry than Southerners. Do people have different media metronomes, too? No doubt, but is the difference as stark as the difference with highway driving?

Rest assured, some people view the media as something with which to "curl up" in front of a warm fireplace. Old media like books, magazines, newspapers, broadcast TV, and radio have a powerful attraction to consumers who seek comfort and relaxation. That there are many in this traditional audience who are older has gradually increased the average age of network audiences.[7] But the overall trend in a new media environment is toward less comfortable choices.

Ease of Ad Avoidance

In an advertiser-supported model of media content, the loss of urgency is made worse by the ease of avoiding advertising. Such ad-avoidance, or *advoidance*, is a by-product of rapid diffusion of DVRs in the home, often bundled with set-top boxes. In addition, automatic ad blockers skip over the ads on the Internet. Once limited to pop-up blockers, these ad killers on Web browsers (e.g., IE7Pro) prevent display-type advertising entirely.

Another type of advoidance is simply spending less time with ad-based media. An iPod loaded with music displaces an hour per day of time spent listening to the radio.[8] One-dollar DVD rentals displace time spent viewing commercial television. Libraries of TV-show DVDs are an attractive alternative to "57 channels (and nothin' on)," to borrow the title of an old Bruce Springsteen song.

Competing Timekillers

Traditional media no longer have a monopoly on leisure time in the home. Even after the arrival of new choices via cable, the television set occupied the lion's share of people's home-based recreation. Rubin measured the propensity of some media consumers to find avoidance of boredom as a use and/or gratification of watching television.[9] Contrasted with "instrumental" uses, the *habitual* uses of the media have now been studied for decades.

Thus, many media uses can be viewed as killing time, i.e., nothing better to do. In the 1990s, the rise of the World Wide Web as a means to easily surf the Internet has resulted in a new competitor for wasting time. In 2007, a fledgling social networking medium, Facebook, began a rapid growth in user popularity, from 30 million then to 300 million users in 2009.[10] Originally the domain of college students, the Facebook website is another in a series of online timekillers that compete with traditional media. Even older media users now join Facebook to keep up with their youthful relatives. Not many young parents mail photos to their parents. It is easier to post a snapshot or video clip on Facebook or YouTube. Online videos that went "viral" via e-mail had begun spreading on Facebook and MySpace by 2009, the year that Facebook alone accounted for 20 percent of all online advertising impressions.[11]

Very young children have come to view traditional media as an alternative to the Internet, rather than the Web being an alternative to television.[12] My own pre-teen children spend most of their media time on YouTube and video games, exposed to almost very few commercials within advertising-based media. They do not plan their media consumption; they expect all of it to be accessible whenever they want it. They know nothing of old and new media, only the TiVo and the computer. Just as the social media habits of young people infiltrate older audiences, so too will the use of online video as an alternative to traditional schedule-based television or either cable or broadcast channels. In 2009, about 75 percent of Millennials (ages 14 to 25) reported that the computer is more of an entertainment device than their television set.[13] As this population ages, it is doubtful that their media use patterns will transform somehow to view television as a more dominant force. The fuse has been lit and the broadcast networks may not escape the ticking time bomb.

Abundant Choice

Disruptive digital technologies have altered broadcast media and their pre-1980s business models based on few competitors, captive audiences, and unavoidable advertising. Thanks to technology, the big analog media are threatened. Consumers will ad-block the Internet and viewers will use TiVo to skip 50 to 60 percent of the ads. Product placement will only go so far, because not every product or service is readily inserted into a storyline. For the audience, the ultimate "control" may be skipping the advertising interruptions, but so, too, is access to more content in more locations.

The number of *choices* is primarily to blame for the decline in network advertising's value, even before ad-skipping. Broadcast television started dying in the 1980s when cable/satellite began producing fresh content available 24/7 instead of being relegated to some daypart ghetto. Advertisers so desperately *need* mass audiences, but the simple truth is that, short of mega-events, too many diversions are competing for too few eyeballs for anyone to make enough money.

Some new choices come from new technologies. MP3 downloading was the beginning of the end for over-the-air radio. Young kids do not want radio. Radio is skipping an entire generation. The picture for television networks is similar. For the first time, people ages 18-24 spent nearly the same amount of time in 2009—roughly five hours—watching Internet video each month as they did watching television programs. Other age brackets watched half as much or less Internet video.[14]

Benjamin Bates notes that access, choice, and control are three coveted aspects of the modern viewing landscape.[15] Viewers are willing to pay extra for programming that offers more of these three aspects (e.g., cable, DBS, satellite radio). Bates has called for "a new way of thinking about media products and markets," which may indeed create problems for older models.

Could it be that content is no longer king? Mark-to-market accounting and bankers created the problem with media company valuations. In good times, accountants were free to label created content as a long-term asset. In bad times, such assets are often restricted to today's value rather than any future value. It is difficult to determine how content will be re-valued as the economy recovers from the Great Recession. The newspaper and magazine industries may not be able to get any of their value back. Some of the more valuable content, like expensive Hollywood movies, may only make a great deal of money in theaters.[16]

Free Content

The expectation of free content is the final nail in the coffin of old media business models. Examples of free content replacing paid content are abundant. Wikipedia is the set of encyclopedias that consumers formerly bought from Encyclopedia Britannica.[17] Google Docs is the expensive Office Suite you formerly bought from Microsoft. Even the lucrative market for online pornography has dried up with the advent of free Internet adult video websites and filter-free image searching. Ask Larry Flynt, who in 2009 was looking for Federal aid for the near-bankrupt pornography industry. The "no pay, no content" world no longer applied.

Free content is often a ploy to build a loyal following that can be converted to paid subscribers, once users see the benefits of the content. The problem is, someone else comes along near the upward arc of popularity and also decides to compete. The first provider cannot sell the content because the second provider starts giving it away, and so on, ad infinitum.[18] In an era of scarcity, the providers can at least sell advertising while building a subscriber model. In an era of abundance supply, however, the model changes.

Even Rupert Murdoch has had trouble selling *Wall Street Journal* content, which as recently as 2005 was the shining example of content that would always attract paid subscribers. But the Internet is just too big and has too many choices for anyone to get rich anymore. The "long tail" was an idea that sold a lot of books, but has not always been a successful business model.[19]

As network advertising continues to erode, executives plan to take content "everywhere" but not following the fewer-interruptions model made popular by sites like Hulu. In 2009, broadcast networks began announcing their plans to stop giving away their content online. Unfortunately, online users are likely to resist the change. At least one study finds that over half of the online audience will refuse to pay for content.[20]

True, the demand for sitcoms and movies will continue. Moviegoers will still shell out big bucks to sit in the dark, for a while at least. But the economics of popular TV shows is shaky now, because the size of the competing channels and sheer number of competitors makes it difficult to get rich owning content. *Seinfeld,* for example, worked profitably because the networks still captured enough share of the audience to be patient through the first and second meager-viewing seasons. It had time to catch on. One wonders if a show about nothing would last more than six weeks in today's quick-hit world.

The Counterarguments

Given the trajectory of the argument that loss of urgency, ease of advoidance, competing timekillers, and abundant (often free) content has impaired broadcasters' fabled license to print money, this is perhaps the perfect place to introduce the cold water that regularly gets splashed onto the idea that anything is wildly amiss in the world of mass media and its support by advertising and subscriptions. Each can be grouped by the source, either people themselves (P) or advertisers and their agencies/providers (A).

1. (P) Old dogs cannot learn new tricks. People just want effortless entertainment. Yes, they know how to lean forward, but they prefer to lean back.
2. (A) Product placement will permit advertisers to sneak branding mentions into programs.
3. (P) Skipping ads and avoiding messages requires effort. No one wants to press a button all the time to skip messages. Control and choice are antithetical to comfort.
4. (A) Surveys support the notion that audiences think advertising is a fair price to pay for free or cost-subsidized content.
5. (A) These gadgets are mostly limited to young people or those in urban areas; changes for the rest of America are many years away.
6. (A) People who are interested in buying or replacing a product want to see advertising.
7. (P) Shoppers like coupons and sales. Without advertising, they cannot learn about exciting promotions and sales. Many women enjoy the color pre-prints in the Sunday paper.
8. (A) The VCR did not increase ad-skipping, according to most research. The DVR is a glorified VCR. Thus, the DVR is similarly unlikely to cause much ad-skipping. Besides, only a minority of homes have DVRs.

9. (A) Radio is a free medium; it's live, local, and friendlier than a pocket jukebox. The receiver itself is dirt cheap.
10. (A) No one will watch media content on a tiny screen when they have a big flat screen at home.
11. (A) Live events will save the day for old media.
12. (P) Young folks have never been much for reading newspapers, even before the rise of new media. The urge will kick-in when they become homeowners and parents.

Because people retreat to their cocoons in tough economic times, the deniers frequently mention television.[21] Hu, Lodish, and Krieger compared experimental groups of people who either did or did not see TV advertising, and reaffirmed the effectiveness of a previous longitudinal study.[22] But extrapolating experimental subjects to the real world, where presumed exposure is now different than actual exposure to advertising, is another matter.

Neff cited other studies of TV commercial effectiveness without commenting on methodological issues or outright bias.[23] For example, comparing those advertisers who use any TV ads at all to those who use none may mask the real effect, by using nominal level variables instead of more reliable continuous variables. Some of the published studies are sponsored by the organizations with the most to gain by finding no decline in TV advertising effectiveness, so one should be cautious interpreting their imprecise methods.

Another defense of TV advertising claims that adaptive viewers derive more pleasure from interrupted programs. Nelson, Meyvis, and Galak conducted six studies and reported that although people preferred to avoid commercial interruptions, these interruptions actually made programs more enjoyable (study 1), regardless of the quality of the commercial (study 2), even when controlling for the mere presence of the ads (study 3), and regardless of the nature of the interruption (study 4). However, this effect was eliminated for people who are less likely to adapt (study 5), and for programs that do not lead to adaptation (study 6). Their study suggests that adaptability is an important measure.[24]

Some observers could argue that media economics in 2009 was an anomaly of the Great Recession, but the systemic effects that underpin the losses of advertising revenue are more likely to blame. Erik Sass wrote, "The general downward trend was in evidence well before the recession began, reflecting fundamental secular shifts in media consumption and advertising. If anything, the economic downturn is merely accelerating this process."[25] Why has advertising deteriorated during the first decade of the new millennium?

Perhaps, as Shelly Palmer explains: "Advertising sucks."[26] Or as Tony Granger claims: "Advertising agencies are dead."[27]

Advertising itself is not to blame. When it works, advertising is wonderful. If marketing is the engine of commerce, then advertising is the lubricant. In a typical example, the seller has a new or underpromoted product or service. Advertising communicates a message to the buyers, typically part of a mass or targeted audience, with the goal of selling more—or making current customers feel happy or smart, which in turn promotes future sales. The value justifies the expense, but most ads are very expensive. Local television commercials in top 10 markets cost between $4,000 to $45,000 for 30 seconds. The same spot broadcast during network prime time ranges between $80,000 and $600,000. On cable, the same 30 seconds in prime time runs $5,000 to $8,000.[28]

As large as the expense of advertising seems, it is spread over the size of the audience for the message. Commercials are priced on cost per thousand reached (CPM) or cost per ratings point (CPP). The impact is measured on the number of gross impressions (GI). Also quality of the message plays a major role. Stickiness of websites is considered for online ads; for all media, commercials that are engaging bring more success.

The linchpin of most advertising, however, is the *forced* connection between the media content sought by the consumer and the surrounding advertising. Three things can go (and have gone) wrong with the old system of advertising. First, remove the linchpin and the justifying value is diminished. Second, dilute the attention to the message or content with additional choices and the value is similarly reduced. Third, increase the number of competitors with a particular type of communication (print, video, aural) and the value again goes down.

For years, network media executives assumed that consumers felt advertising a fair price for free over-the-air television. Surveys supported that notion. The opinions of mass audiences may have been tied to limited options (i.e., just a handful of broadcast networks): No one cared to pay for limited choices. Nowadays, however, viewers have begun to show far greater preference for paying for their media content, especially if they receive unlimited choices. Accenture, a media consulting company, surveyed 14,000 consumers in 13 countries in early 2009, finding that 49 percent (an annual 12 percent increase) said they would pay a monthly fee for unlimited digital content.[29]

As this chapter focuses on network advertising, the situation for broadcast stations is pretty dire. For example, EBITDA multipliers for the sale of broadcast television stations are down 40 percent (about 7 times cash flow in 2009, which is a 25-year low).[30] True, advertising will live on in some form, but not big enough to feed the media giants, especially the television networks. Ads that are unblockable or unskippable will find continued success, like those commercial messages found on ballpark walls, in gas station videos, and on highway billboards, but the model for traditional mass media is irretrievably broken.

Predictions of the demise of old-school advertising are commonplace. Longtime media observers have been predicting the end of advertising. Bob Garfield began writing a series of articles about "the chaos scenario" as early as 2005. His 2009 book added some prescriptions for the future.[31] Garfield's most compelling argument is that advertising works best in media that are scarce. That scarcity does not reflect the digital world, he notes, where abundance drives down the unit price of marketing messages.

Consumers like product information and they enjoy being aware of innovations and change, but they grow weary of overexposed messages. Granted, repetition is essential to success, but advertising wears out its welcome, usually to the point that consumers will have none of it, if given the chance to avoid the pitch.

Because the cost of network advertising is so high, the plight of the networks is particularly risky. Garfield wrote this in 2007: "Because no other medium offers the reach of TV, advertisers have continued to pay more and more per thousand viewers. But economics will have its due. The law of diminishing returns will eventually prevail. Those who have perennially spent more and more for less and less will finally say, 'No more,' and take their money online whether there is sufficient ad inventory or not."[32]

By "eventually" Garfield apparently meant 2009, when the economic collapse created the perfect storm for television advertising, with double-digit declines.[33] Advertisers started talking about the need for a new way to market to consumers, one that shifted from telling-selling to building relationships. Social networks and viral videos have moved those advertisers closer to their goal of engagement, a buzzword that galvanized the first decade of the new century.[34]

In his 2007 article, Bob Garfield identified five reasons the online world will displace traditional modes of advertising:

1. People don't like ads—The evidence is clear, when somewhere between 50% and 70% of DVR users skip ads.
2. But they crave information about goods and service—Search-engine marketing is better than traditional ads because search is contextual, measurable and information-rich.
3. The consumer is in control—"The fact is, people care deeply— sometimes perversely—about consumer goods. . . . What they don't like is being told what they should care about or when they should be caring."[35]

4. Diversion of ad budgets—Crunching the vast amount of data from interactive marketing will drain billions of dollars from existing ad budgets, unlocking "the very power of aggregation, information, optimization and customer-relationship management that will render most image advertising impotent and superficial."[36]

5. Pay-per-view—If, in the near future, most content is paid for by the user, either via subscription, like HBO, or à la carte, like pay-per-view or iTunes, then advertising would be eliminated from the equation. If micropayments ever become practical, pay-as-you-go would allow users to seamlessly buy, for instance, newspaper content on an edition-by-edition or even article-by-article basis. [Garfield cites Bruce Owen]: "The willingness to pay by consumers is far greater per eyeball than the willingness of advertisers."[37]

If advertising on TV is somehow less ideal for reaching Americans in their homes, then what other kinds of advertising still works? The simple answer is outside the home. Ambush advertising media reach audiences in elevators, public spaces (and along the roads to and from), sports venues, restaurants/bars, grocery stores, and wherever people wait in a line. Out of home advertising (OOH) is often called "the fourth screen" (assuming the first three screens are the home TV, the computer screen, and the cellphone).[38]

Ads that continue to capitalize on low choice, forced exposure, and focused attention will work. For example, place-based ads take advantage of a captive audience, like those in an airplane or, more common, patrons in a movie theater. Katy Bachman noted that "auto companies are increasingly turning to the silver screen as a primary medium [rather than television, which has two to five times weaker ad recall]" for introducing new models.[39] Even standing in line at the concession stand is no escape if existing plans to install digital ad platforms in movie theater lobbies are successful.[40]

Even on network television, one finds plenty of evidence that advertisers on network television want to trap the target audience into viewing marketing messages. Advertisers are testing some novel schemes to enforce the continued attention to ads. For example, programmers have relinquished the lower portion of the screen to any number of ploys to attract eyeballs. These so-called snipes began as reminders for Nielsen diary-keepers who needed help knowing what channel they were viewing, which was a growing problem as new choices mushroomed. Viewers hated these channel bugs, but became acclimated over time. Then the promotional snipes became animated, drawing new ire from viewers, but only for a short time.

Advertisers have been champing to get into the act. As this chapter is being written, the search engine Ask.com has just started an ad campaign that relies mostly on snipes that show up in the lower part of the screen during selected cable programs.[41] If the audience can be slowly persuaded to tolerate these intrusions in the same way they were taught to put up with commercial interruptions, network television could survive, but if audiences' willingness to pay is greater than their willingness to view intrusive advertising, the future of advertising looks dim.

The advertising model, as practiced for the last hundred years or so, has been one based on interruption. It is a consistent tactic across TV/video, radio, and even print. It continues because it works. And, in the digital video arena, pre-stream and mid-stream commercials will continue to exist for some time, because they work to help marketers achieve their goals. However, marketers have also begun to recognize that broadcast networks cannot deliver mass audiences, as cable networks chip away at their viewers.

Can Old Media Adapt?

Not necessarily, because the idea that old media need to adapt to new digital technologies may be very naive. Moving media content from one platform to another is not as easy as it appears. As explained above, the value of old media companies was built upon their relative scarcity, high barriers to entry, expensive content, and exclusive access to viewers who need to fulfill one of the five basic functions of all media (surveillance, interpetation [correlation], transmission of values, linkage, and entertainment).[42]

Many of these five functions privileged one medium over another. Newspapers, unfettered by the FCC, could present many sides of an issue without worrying much about fairness or balance or personal attack. If a reader really enjoyed opinion, the only other alternative was the magazine, which seldom dealt with local matters.

The Internet changed that. Blogging supplanted the opinion columns and many of their best writers. Getting your letter to the editor chosen by the newspaper was sometimes like buying a lottery ticket, but posting a reply to a blog can be done by anyone, without regard to grammar or spelling or even logic. One could argue that the older, refined product was better, but the consumer has the final say.

Gatekeeping and agenda-setting rely on limited numbers of gatekeepers and agenda-setters. Expanding the exclusive club devalues all the members. For example, newspaper monopolies long fulfilled the needs of local merchants by being able to present long lists of items for sale. Radio and TV had a tough time competing with the density of the large display ad, with which auto buyers could examine the cars for sale locally or homemakers could pre-select foodstuffs and household necessities. So once the dominant paper, or two, clawed its way to the top, the revenue stream really had no competition.

Likewise, classified advertising had no equal and few alternatives. A person could post a note card at the laundromat or grocery bulletin board, but the classifieds reached everyone, although the smart newspapers ran find-your-name contests to get readers to view the small-print columns of ads. If you wanted to sell your car, you could put a sign in the window, or even park it at a high-traffic location, but the quickest way to selling was a classified ad. Newspapers soon came to rely on 40 percent to 45 percent of their revenue from classifieds. It was like fishing in a barrel. Even the free trader-ad weeklies fell short of the reach of a daily newspaper.

Then craigslist.org and eBay.com changed all that. Newspapers began to see their revenue from classifieds plummet. Some dailies began offering free ads for items under $100, simply to attract return customers for higher ticket items. Age differences came into play. Older readers like the comfort of old ways. Young consumers (not exactly readers) found convenience in selling items on craigslist.org, because they were already spending huge chunks of time online, tweaking their Facebook page and following their friends on Twitter.

The abundance of choice in mass media in the twenty-first century is a curse on old media, but even a few choices can be overturned by just one additional choice. The following food/restaurant metaphor of media content providers has appeared previously in my writings.[43] Different programming competes in the same way different restaurant specialties compete: sitcom versus drama is akin to pizza buffet versus steakhouse. If a town has only one restaurant or three restaurants, then people learn to enjoy their visits regardless. Eating out once in a while is preferable to standing over a hot stove every day of the week. It does not really matter if all three restaurants are fantastic, however. As soon as another eatery comes to town, people will flock to it. The new restaurant has terrific business until another new restaurant comes along, or the patrons tire of the Thai menu.

Audiences love new choices. This was learned first-hand when a fourth television station arrived in Ft. Wayne, Indiana, in the early 1980s. It did not matter whether the three network affiliates in those pre-Fox days of prime-time network television were doing a spectacular or mediocre job of serving the public. The viewers who could pull in the new signal, filled with old movies or old syndicated off-network shows, were thrilled. This phenomenon had an impact on their viewpoint regarding quality of service. Choice trumps quality, especially where viewing venues are already limited. The latest media landscape is all about choice and control, both of which network broadcasters fail to offer with their limited menu of choices.

The Future of Television

The year 2009 was a very bad year for broadcast television, as noted earlier. Indeed, Dave Morgan has predicted a "cross-platform video future" and poses questions about the existing television industry's plans to launch TV Everywhere (where subscribers would receive video content whenever/wherever they want).[44] These questions include the following: who will control the interface, the content, the revenue, the packaging, and the authentication of paid users? According to Time Warner CEO Jeff Bewkes, "If you want to watch your favorite TV network or shows through broadband on any device—PCs or mobile—you can do it as long as you subscribe to any multichannel provider. . . It's a natural extension of the existing model."[45] Presumably the viewer will be trapped into seeing the commercials, at least until someone devises a portable DVR device that stores the program onto hard-disk storage where ad-skipping might still occur. The designers of TV Everywhere promise to turn off fast-forwarding, but it is unclear how they would accomplish this feat, or thwart other designers who might invent a scheme to circumvent the networks' attempts to keep advertising a *forced* choice.

Diane Mermigas paints a very bleak picture of old-style broadcast television: "TV stations' ability to excel in the nascent but promising world of hyperlocal information and services is hindered by a slew of uncontrollable forces. There is the collapse of core ad categories, such as automotives, which has contributed about one-fourth of all TV station revenues and will never fully recover. Internet-connected streaming video for PCs and mobile devices will continue to minimize and fragment television. Despite massive reductions in workforce and legacy operations, the pooling of local news-gathering and ad sales resources, and a growing Web presence, TV stations' economic quandary increasingly mirrors that of declining newspapers."[46] Adding to the trouble is the influence of Internet-enabled television receivers arriving in the electronics store showrooms by 2010.

Broadcast television is eager to survive, but so far Web ads are not making much money for stations. Ben Fritz writes "networks may actually be undercutting themselves in their quest to avoid the fate of the music business."[47] Competition from online distribution could make the Internet the common enemy that finally unifies broadcasters with cable channels. This may explain the eagerness of broadcasters and cable channels alike to pursue TV Everywhere, as a means to thwart those who are using their broadband connection as a subsitution for scheduled programming. Moreover, broadcast networks by 2009 were openly discussing the possibility of becoming cable networks.[48] If you can't beat 'em, join 'em.

At this writing, a sliver of hope for broadcast stations comes from the Open Mobile Video Coalition (OMVC), a group of over 60 stations in 22 cities that planned to transmit news, entertainment and sports to portable devices (to be available by Christmas 2009).[49] Ironically, this initiative comes at the same time that the FCC was considering a move to reclaim broadcast spectrum to be used for mobile carriers. The cost per station is only $250,000 for a new transmitter exciter, but the plan may prove "too little, too late" for broadcasters.

Predictions

Making predictions about the future of the media is easy. And fun. The problem is that eventually glib descriptions are judged by their predictive power. This final section of the chapter is not a first gaze into a crystal ball. In 1997 the media world of the future was imagined in a chapter in a book on broadcast television.[50] At that time, this author's tea leaves called for distribution to prevail over content, foreseeing that new technologies would change everything. This author had no idea how accurate the "content is not king" prediction would hold, ten years later.

About the same time, this author had an opportunity to review other predictions of the future of broadcast television that held little face validity.[51] Ever since, it has become obvious that writers can make compelling and interesting predictions that nothing much is really going to change, that media will adapt and get better, that the future is rosy, and blah, blah, blah. These authors may even draw *more* attention to their predictions by being so counterintuitive in the face of seismic shifts among the media.

At the risk of being wrong, here is this author's new (and very brief) prediction for the future of television: Ten years from now, when people are watching video on headset screens and earpieces, linked to cellphones, young people will marvel that older folk ever stared across the room at a screen. And followed a schedule. And watched commercials.

Notes

1. Raymond Bauer, "The Obstinate Audience: The Influence Process from the Point of View of Social Communication," *American Psychologist* 19 (1964): 319–328.

2. Bob Garfield, *The Chaos Scenario* (Nashville: Stielstra Publishing, 2009).

3. Saul Berman, Bill Battino, Louisa Shipnuck and Andreas Neus, "The End of Advertising as We Know It," IBM Global Business Services (2007), http://www-03.ibm .com/industries/global/files/media_ibv_advertisingv2.pdf (30 October 2009).

4. Wayne Friedman, "TargetCast: 3Q Network Ads Drop 16%," *MediaDailyNews*, October 28, 2009, http://www.mediapost.com/publications/?fa=Articles .showArticle&art_aid=116241 (30 October 2009).

5. Alan Rubin, "Television Uses and Gratifications: The Interactions of Viewing Patterns and Motivations," *Journal of Broadcasting* 27(1983): 37–51.

6. Michael Schneider, "DVR Retunes Television," *Variety*, September 29, 2009.

7. Gary Levin, "Network Audiences are Showing Their Age," *USA Today*, June 27, 2007.

8. Douglas Ferguson, Clark Greer, and Michael Reardon, "Uses and Gratifications of MP3 Players among College Students: Are iPods More Popular than Radio?" *Journal of Radio Studies* 14(2007): 102–121.

9. Ferguson, Greer, and Reardon, 2007.

10. Caroline McCarthy, "Facebook Hits 300 Million Users," *CBS News Tech*, September 15, 2009, http://www.cbsnews.com/stories/2009/09/15/tech/cnettechnews/ main5313658.shtml (30 October 2009).

11. "Inside Facebook," September 2, 2009, http://www.insidefacebook.com/ 2009/09/02/social-networks-more-than-20-percent-of-online-display-ads/ (30 October 2009).

12. Douglas Ferguson and Elizabeth Perse, "The World Wide Web as a Functional Alternative to Television," *Journal of Broadcasting & Electronic Media* 44(2000): 155–174.

13. John Loechner, "Some Communications Segments Forecast Up, Traditional Down," *MediaDailyNews*, March 4, 2009, http://www.mediapost.com/publications/ ?fa=Articles.showArticle&art_aid=101235 (30 October 2009).

14. Elizabeth Holmes, "Mobile, DVR Video Log Fastest Growth," *Wall Street Journal* (Eastern Edition), February 23, 2009.

15. Benjamin Bates, "Transforming Information Markets: Implications of the Digital Network Economy," *Proceedings of the American Society for Information Science and Technology* 45(2009): 11.

16. Douglas McIntyre, "Content, Once King, Becomes a Pauper," *Time Business & Technology*, February 11, 2009, http://www.time.com/time/business/ article/0,8599,1878711,00.html (30 October 2009).

17. "Although the Reliability Was Suspect at First, Wikipedia Has Solidified Its Respectability since 2006." http://en.wikipedia.org/wiki/Reliability_of_Wikipedia (30 October 2009).

18. A possible exception to this trend is Hulu, launched by NBC and Fox in 2007. Hulu announced in 2009 its plans to become a pay service by 2010.

19. Anita Elberse, "The Long Tail Debate: A Response to Chris Anderson," *Harvard Business Blog*, comment posted July 2, 2008, http://blogs.harvardbusiness.org/ cs/2008/07/the_long_tail_debate_a_respons.html (24 October 2009).

20. Erik Sass, "Net Loss: Most Online Readers Won't Pay For Content," *MediaDailyNews*, October 27, 2009, http://www.mediapost.com/publications/?fa=Articles. showArticle&art_aid=116242 (30 October 2009).

21. For example, Jack Neff, "Guess Which Medium is as Effective as Ever: TV," *Advertising Age*, February 23, 2009.

22. Ye Hu, Leonard Lodish, and Abba Krieger, "An Analysis of Real World TV Advertising Tests: A 15-Year Update," *Journal of Advertising Research* 47(2007), 341–353.

23. Hu, Lodish, and Krieger, 2007.

24. Leif Nelson, Tom Meyvis, and Jeff Galak, "Enhancing the Television-Viewing Experience through Commercial Interruptions," *Journal of Consumer Research* 36 (2009), http://papers.ssrn.com/sol3/papers.cfm?abstract_id=1007767 (30 October 2009).

25. Erik Sass, "It Was A Very Bad Year: Old Media Suffers Long-Term Declines," *MediaDailyNews*, April 1, 2009, http://www.mediapost.com/publications/?fa=Articles.showArticle&art_aid=103327 (30 October 2009).

26. Marisa Guthrie, "Cable Show 2009: Panel Questions Whether Advertising 'Sucks'," *Multichannel News*, April 2, 2009, http://www.multichannel.com/article/191212-Cable_Show_2009_Panel_Questions_Whether_Advertising_Sucks_.php?rssid=20060 (30 October 2009).

27. Becky Ebenkamp, "CLIO: 'Advertising Agencies Are Dead,' Says Exec," *Brandweek*, May 14, 2009, http://www.mediaweek.com/mw/content_display/esearch/e3i3e-5aa5e0b30aa48ef7eaf09462e6a04a (30 October 2009).

28. "Comparison of Advertising Costs," n.d., http://www.polepositionmarketing.com/library/advertising-comparison.php (30 October 2009).

29. Steve McClellan, "Viewers Will Pay to Go Adless," *Adweek*, April 20, 2009, http://www.adweek.com/aw/content_display/news/media/e3if26a27fe344b20e4b8ae25db261619ff (30 October 2009).

30. Harry Jessel, "Bankruptcy Cloud Looms Over TV Owners," *TVNewsCheck*, April 20, 2009, http://www.tvnewscheck.com/articles/2009/04/20/daily.31/ (30 October 2009).

31. Jessel, 2009.

32. Garfield, "The Post Advertising Age," *Advertising Age*, March 26, 2007, http://adage.com/article?article_id=115712 (30 October 2009).

33. Friedman, 2009.

34. Leland Harden and Bob Heyman, *Digital Engagement: Internet Marketing That Captures Customers and Builds Intense Brand Loyalty.* (New York: AMACOM, 2009).

35. Garfield, 2007.

36. Garfield, 2007.

37. Garfield, 2007.

38. Some observers refer to cell phones as the 4th screen, after movie screens. See also http://www.ovab.org/ (30 October 2009).

39. Katy Bachman, "Cinema Sellers Attract Campaigns From Nissan, Kia," *Mediaweek*, February 1, 2009. http://www.mediaweek.com/mw/content_display/esearch/e3i3b5ee64b200b60e1fea83011a4e5c152?pn=1 (30 October 2009).

40. Erik Sass, "Screenvision, SeeSaw Expand Networks," *MediaDailyNews*, April 3, 2009. http://www.mediapost.com/publications/index.cfm?fa=Articles.showArticle&art_aid=103477 (30 October 2009).

41. Brian Steinberg, "Ad Skippers Beware: Ask.com Going After You with TV Crawl," *Advertising Age*, March 2, 2009, 4–33.

42. Harold Lasswell, *The Structure and Function of Communication and Society: The Communication of Ideas.* New York: Institute for Religious and Social Studies (1948), 203-243; Charles R. Wright, "Functional Analysis and Mass Communication," *Public Opinion Quarterly* 24(1960): 610–613.

43. Douglas Ferguson, "A Framework for Programming Strategies," in *Media Programming*, ed. Susan Eastman and Douglas Ferguson (Boston: Wadsworth, 2009), 4.

44. Dave Morgan, "Who Will Own 'TV Everywhere'?" *Online Spin*, October 15, 2009, http://www.mediapost.com/publications/?fa=Articles.showArticle&art_aid=115551 (30 October 2009).

45. Michael Learmonth and Andrew Hampp. "TV Everywhere—As Long as You Pay for It," *Advertising Age*, March 2, 2009, 1–32.

46. Diane Mermigas, "TV Station Revenue Crisis: Mind the Gap," *MediaPost* blog, July 6, 2009, http://www.mediapost.com/publications/?fa=Articles.showArticle&art_aid=109139 (30 October 2009).

47. Ben Fritz, "TV Goes Overboard with Internet," *Variety*, March 13, 2009, http://www.variety.com/article/VR1118001234 (30 October 2009).

48. Sam Schechner and Rebecca Dana, "Local TV Stations Face a Fuzzy Future," *Wall Street Journal* (Eastern Edition), February 10, 2009.

49. Glen Dickson, "ATSC Eyes Digital TV's Future," *Broadcasting & Cable*, December 8, 2008, http://www.broadcastingcable.com/article/160355-ATSC_Eyes_Digital_TV_s_Future.php (30 October 2009).

50. Douglas Ferguson, "The Future of Television," in *The Broadcast Television Industry*, ed. James R. Walker and Douglas A. Ferguson, (Boston: Allyn & Bacon, 1998), 185–199.

51. Douglas Ferguson, "Book Review and Criticism: *Television Today and Tomorrow* by Jankowski & Fuchs," *Journal of Broadcasting & Electronic Media*, 40(1996): 145–148.

5

Going Viral

Mass Media Meets Innovation

Joan Van Tassel

THIS CHAPTER COVERS MEDIATED VIRAL communication (MVC). It occurs when hundreds, thousands, even millions of people pass along a message to others, spreading it using a media technology like the telephone or the Internet. The chapter covers (1) the role of the Internet and the Millennial generation in making viral communication faster and more far-reaching than it has been in the past, when it depended only on a combination of mass media transmission and interpersonal conversations; (2) introduces classic models of interpersonal and mass media communication and shows how MVC extends those models by adding a feedforward path to them; (3) identifies the stages of the feedforward path as origination or exposure, uptake, replication or re-expression, and distribution; (4) identifies the characteristics of feedforwarded content as replicable, memorable, memetic, personal, incomplete, imperfect, and direct; as well as, (5) shows how adding a feedforward path aligns the spread of media messages with diffusion of innovations processes.

The chapter also discusses how professional communicators and media managers can harness MVC to fulfill their objectives. It identifies the skills they will need to employ to carry out viral communication efforts. Finally, it suggests possible business models they can adopt to support such campaigns.

Media Virus: Have You Been Infected?

Human beings communicate to learn, laugh, love—indeed, to live. It is believed that speech in *Homo sapiens* evolved about 200,000 years ago.[1] Viral communication has existed as long as people have exchanged messages: The entire aim of communicating at all is to pass a message from one person to another and, once a message is disseminated, people can continue to pass on information that is perceived as important or interesting. While no records survive from tens of thousands of years ago, it is not difficult to imagine a lookout shouting: "Our enemies are coming!" and that warning spreading quickly to an entire village.

Since then, people have exchanged information in person, face-to-face and at a distance using technology. They have used smoke signals, fire beacons, drums, the heliograph, semaphore, maritime flags, and complex telecommunication networks.[2] Today, people receive messages from family, friends, co-workers and acquaintances, the mass media, and online contacts.

The spread of contagious disease from one person to another through contact is one that nearly every human being has either experienced or heard about. As people also spread information through contact, it is easy to see why the metaphor of contagion has exerted a powerful sway over communication scholars, going back to the 1930s.[3] The rise of the mass media diverted research attention from direct message exchange toward the effects of mediated communication. However, now the Internet has given new life to contagion models, because it makes it possible for all kinds of messages, including those transmitted from the mass media megaphone, to be carried far and wide through direct personal contacts, reaching an enormous number of people via online technologies.

As science has learned more about viruses, the viral metaphor has proven to adapt well to the media environment, leading Douglas Rushkoff (1994)[4] to coin the term "media virus." Consider the behavior of biological viruses: They are transmitted through direct contact or through the air. They select specific hosts into which they insinuate themselves, where they reside, often without producing symptoms. However, the host responds. Although viruses are unable to replicate on their own, they are parasites that take over and subvert the resources and processes of the host to replicate themselves. The means of viral (re)production are genetic: DNA or RNA. The worst viral infections can cause death to the host.

Like biological viruses, media viruses take up residence in a host; however, the host is self-selected, rather than being specified by the virus. The host receives the virus from a mass media source over the air (wireless transmission of television, radio, the Internet, or mobile telephone signals), over a wire (cable or telephone company), or through a person-to-person interaction. If infected, the host initially shows no symptoms, but over time may begin to replicate the media virus by expressing or repeating it to interpersonal contacts in person or via the telephone and Internet messages. The model of

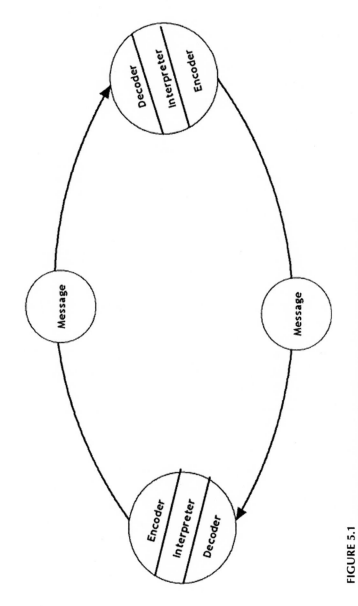

FIGURE 5.1
One-to-One: The Schramm-Osgood Model of Interpersonal Communication

interpersonal communication, presented in figure 5.1, shows the exchange of person-to-person messages. (A model is a simpler representation of a complex structure or process, clarifying the main parts and excluding less important parts.)

The means of spreading a media virus's (re)production are cultural, through *memes*. A meme was defined by biologist Richard Dawkins[5] as a unit of cultural transmission. He conceived of a cultural equivalent of a gene, a cultural entity that replicates itself and evolves by residing in the human mind and replicating itself through communication, evolving in the process.[6]

According to Richard Dawkins:

> Examples of memes are tunes, ideas, catch-phrases, clothes fashions, ways of making pots or of building arches. Just as genes propagate themselves in the gene pool by leaping from body to body via sperms or eggs, so memes propagate themselves in the meme pool by leaping from brain to brain via a process which, in the broad sense, can be called imitation. . . . It looks as though meme transmission is subject to continuous mutation, and also to blending."[7]

In true memetic fashion, the meme of media viruses has gradually replicated across the Internet and mutated into "viral media," "viral video," and "viral marketing." These words are now commonly used to describe well-known communication processes. However popular, not all communication scholars think that the concept of memes has value in communication research. Bruce Edmonds[8] and Scott Atran[9] argue that the phenomenon described as a meme is covered by other evolutionary approaches to culture, that it adds nothing new to the understanding of such phenomena, and that there is a lack of substantive results reported by communication researchers. However, Matt Gers[10] defends the analytical usefulness of memes and concludes that they provide explanatory power in the study of cultural evolution.

The Technologies of MVC

Since the introduction of the concepts of media viruses and memes, the technology to spread them has changed a great deal. Communication technologies encourage and allow some forms and types of messaging more than others. Before the Internet, the two modes of communication were interpersonal exchanges and mass media transmission and reception.

Interpersonal communication is interactive. The interaction may be rapid, as with face-to-face or telephone conversations, or slow, as with postal mail. Interpersonal communication affords message flow modalities between individuals and small groups that include one-to-one, one-to-few, and few-to-few exchanges. It does not allow the dissemination of messages to a huge number of people, however: Without some kind of communication technology, a single person could reach several thousand people at most—never several million.

Traditional mass media—television, radio, newspapers, and magazines—are minimally interactive, sending identical messages to all receivers. They offer only limited opportunities for exchange, such as a letter from an audience member. Mass media afford few-to-one and many-to-one message flow modalities between one or a few people and a potentially huge audience, but not a conversation between friends, which depends on an interactive exchange of unique messages.

However, there has always been an important link between interpersonal and mass communication. Thus, although Schramm's classic model of mass media[11] focused on the transmission of messages via mass media, it also formally acknowledged the spread of these messages by active audience members to others in their interpersonal networks, as shown in figure 5.2. The model shows people receiving content from the media, then discussing it with family, friends, co-workers, and casual acquaintances. This word-of-mouth flow

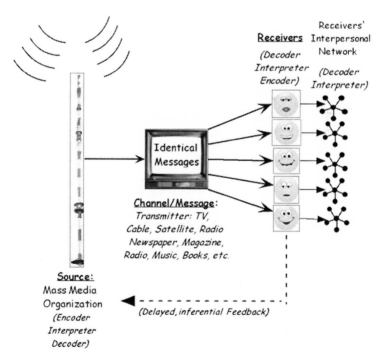

FIGURE 5.2
One-to-Many/Few-to-Many: Schramm's Model of Mass Communication

from the mass media to the audience and then to others was the subject of early communication research and became the basis of the theory of the two-step flow of information and influence.[12]

The Schramm model also incorporates a *feedback loop* that goes from the audience back to the mass media organization that distributed the content. This feedback is characterized as delayed and inferential, meaning that it takes time for audience members to react and express an opinion via letters or ratings research. The spread of mass media messages from active audience members to others in their interpersonal networks was well-described by researchers at the time.

The forward flow of information was not given a name or label in Schramm's model. However, now it has become a more prominent pathway because the Internet provides an efficient vehicle for it. Thus, this chapter proposes to call the forward flow of information the *feedforward path.*

Prior to the widespread use of the Internet, people's interpersonal network was limited by the number of routine face-to-face contacts, telephone calls, and written messages in which they could engage in a day, week, or even a month. Using the Internet, users can send out messages to hundreds of people with the click or a keystroke, as often as they like, sometimes several times a day. The Internet permits a wider variety of communication flow types than any other medium, as shown in table 5.1, based on the Reardon and Rogers[13] analytical categories.

Communication networks like the telephone and the Internet create a surprising number of possible connections between people in the network. It is a phenomenon called Metcalfe's Law, which explained *network effects,* as shown in figure 5.3. Metcalfe's Law states that the value of a telecommunications network is proportional to the square of the number of users of the system (n^2). First formulated by Robert Metcalfe writing about Ethernet networks, it describes the exponential growth of connections in a network.

To summarize, the Internet is a multi-modality communication technology that allows all possible message flows. It is the only media platform that provides for many-to-many messaging, a key change that facilitates the rapidity and breadth of the transmission of media viruses. As a global media platform, the Internet expands the kind of content people can exchange (text, audio, video, interactive) as well as increasing the repertoire of message flows. Moreover, it extends viral communication beyond synchronous person-to-person exchanges to an enormous and growing number of users online. Messages are transmitted along feedforward paths enabled by the Internet, including mass e-mail and Web 2.0 services such as SNS, social bookmarks, instant messenger, and Net-to-mobile texting services. Whether users discover a current meme by watching a TV show, through e-mail from a friend, on a blog, a

TABLE 5.1
Comparison of Modalities of Communication Flow

	Interpersonal	*Mass Media*	*Internet*
Message flow	One to one One to few Few to few	One to many Few to many	One to one One to few One to many Few to one Few to few Few to many Many to many
Source			
Knowledge of audience	Detailed	Scant	Scant
Audience segmentation	Personal, direct knowledge or indirectly through personal contacts	Demographics Psychographics/ Lifestyle Behavioral	Demographics Psychographics/ Lifestyle Behavioral Person-provided (opt-in) Micro-segmentation via multiple databases
Interactivity	Yes	No	Yes
Feedback	Immediate, reciprocal, direct feedback	Delayed, receiver to source, inferential feedback	Immediate, reciprocal, direct or inferential feedback
Asynchronicity	No	Yes, with DVR	Yes
Emotional vs Task-related content	Both	Emotional	Both
Privacy	Yes	Yes	No

Facebook page, or a peer-to-peer database, the Internet is an infrastructure that enables the spread of memes to an enormous number of people around the world, in near real-time.

Karakas[14] calls the Internet a digital ecosystem, *World 2.0.* This "interactive, hyper-connected, immersive, virtual . . . meta-platform" has brought about cultural shifts toward creativity, connectivity, collaboration, convergence, and community. Underlying all of these activities are three characteristics of the Internet: connectivity, interactivity, and meta-media. They may also generate messages that initiate the spread of viral media via the feedforward path, as shown in figure 5.4.

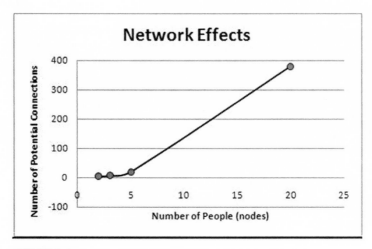

FIGURE 5.3
Exponential Connectivity of Networks

This section examined the role of the Internet in the increasing attention paid to MVC. Two closely related characteristics of the Internet make it an effective purveyor of MVC: Connectivity and interactivity. The next section looks at these elements in more detail to understand who is connected and how.

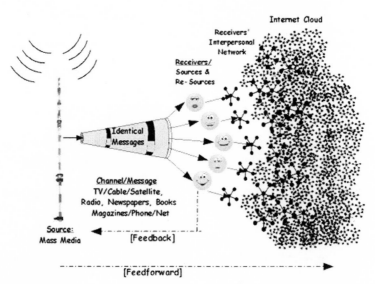

FIGURE 5.4
Mass Media and the Feedforward Path

Connectivity

From a technical standpoint, connectivity means linking devices so that they may exchange data.[15] From the perspective of human communication, it means a communication infrastructure that allows persons to exchange information. Figure 5.4 shows the growth of the Internet, which enables these exchanges. All the communication modalities enabled by the Internet—one-to-many, few-to-many, etc.—depend on connectivity.

There is no question but that one of the most important aspects of the Internet is its immense global reach—remember those network effects. The growth of the Internet, as shown in Figure 5.5, has been extraordinary. However, the quality of connectivity also matters to viral communications, which

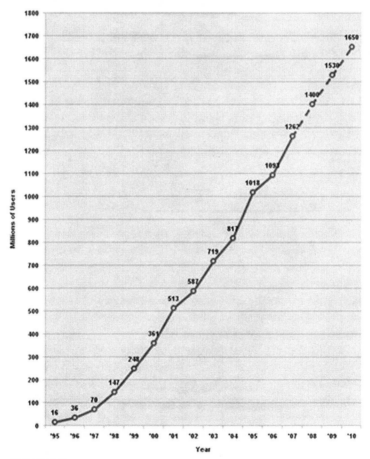

FIGURE 5.5
Growth of the Internet

is expressed in terms of *bandwidth*. Bandwidth is a measure of the amount and speed of communication flow: The greater the bandwidth, the larger the number of data bits can move to and from end users, and the faster they move.

With dial-up service that provides only narrow bandwidth of 56 kilobits (56,000 bits) per second, users can receive text messages easily and download songs with some patience and wait time. Video files are difficult to stream or download, particularly if they are longer than a minute or two. With broadband service that provides three megabits (3,000,000 bits) per second or more of bandwidth, large file types, such as video, can be downloaded, streamed, and forwarded quickly and easily. Even e-mail accounts that limit the size of attachments are not much of a barrier to passing large files along to other persons: "Internet cloud" online services, such as http://www.box.net and other free storage sites, allow users to upload the file and send a link to others, who can then download the files.

Interpersonal connection depends on nearness and synchronous presence in time. One of the great benefits of the Internet is that people can be separated by both space and time, yet still communicate in all ways except synchronous virtual tele-presence. They can exchange text, audio, and video messages, as well as all types of attached files and documents. If they are separated only by space, they can engage in real-time conversational exchanges in text, audio chat rooms and voice-over-Internet (VOIP) services, and video chat rooms and video conferencing services.

Interactivity

Communicative interaction is inherent to relationships; some might go so far as to say that it *is* the relationship. In late 2009, a meme circulated on the Internet claiming that one in eight marriages in the United States in the past year began with the pair meeting online.[16] While there is no proof of the accuracy of the claim, the widespread acceptance of the factoid points to an important insight: The Internet is understood to support the initiation, development, and maintenance of relationships.

To examine the nature of relationships supported by social networking sites (SNS), Ellison et al.[17] studied university students using Facebook. They found that the students connected more with people they knew offline than those they had met online. They separated three types of connections, and found that weak ties were particularly implicated in the widespread forwarding of information: *strong ties* to friends and family, upon whom users depend for emotional support; *weak ties* to acquaintances, which facilitate information diffusion to external relationship clusters; and, *maintenance ties* to keep in touch with people users had known in one stage of their lives, as they progressed through life changes.

Other research has underlined the importance of weak ties in information diffusion.[18,19] Donath and Boyd note that: "Weak ties, the kinds of ties that exist among people one knows in a specific and limited context, are good sources for novel information. Such ties often bridge disparate clusters, providing one with access to new knowledge. Weak ties can be less costly to maintain, and a person who has many weak yet heterogeneous ties has access to a wide range of information and opportunities."[20]

Taken together, these findings suggest that the Internet provides affordances for several types of social relationships, all of which may involve message exchange, including forwarding content or making content available to others via posts, recommendations, and social bookmarking. Because of the low cost and relatively small personal effort, users access the Internet and pass information to more weakly tied interpersonal relationships than they could without Internet technologies. Thus, there is some evidence that the Internet, particularly with the emergence of social media applications, facilitates MVC.

According to the Pew Internet & American Life Project,[21] people from 12 to 32 years of age are the heaviest social media users:

- 65 percent of teens (12-17)
- 67 percent of Gen Y (18-32)
- 36 percent of Gen X (33-44)
- 20 percent of younger boomers (45-54)
- 9 percent of older boomers (55-63)
- 11 percent of silent generation (64-72)
- 4 percent of G. I. generation (73+)</bl>

In light of the potential importance of weak ties to MVC, the notably high use of SNS among teens and Gen Y (sometimes called Millennials) takes on added significance because habits adopted in youth may well carry over into adulthood. The popularity of Web 2.0 among teens and younger adults may also help explain why the content of so much viral communication includes humor, sex, and violence, since these types of messages traditionally have greater currency with young audiences.

And So It Goes: Models of MVC

Theorists and researchers have proposed many differing models of MVC. Since MVC occurs over time, most models incorporate the chronological nature of the process, breaking it into discrete stages of related activities.[22] Some models look at the actions of individuals; others consider how MVC occurs as the aggregate of activities of many people.

This section begins by looking at how individuals participate in MVC. Earlier in the chapter, figure 5.3 showed how the MVC process could begin with a message generated by the mass media. Figure 5.6 details the actions individuals take to spread messages.

Origination/Exposure: Anyone can make content, whether the person is a professional communicator or a casual Internet user. Technorati indexed 133 million blogs between 2002 and January, 2009. According to eMarketer, of the nearly 89 million U.S. content creators, 79 million participate on social networks, nearly 24 million write blogs, and more than 18 million create and upload user-generated video.

After the content is created, it is transmitted to others and, by design or by luck, others are exposed to it. Attention is valuable, and exposure may be quite brief, even cursory. Internet users receive an enormous amount of information. In 2008, Nielsen reported that users accessed the Internet on

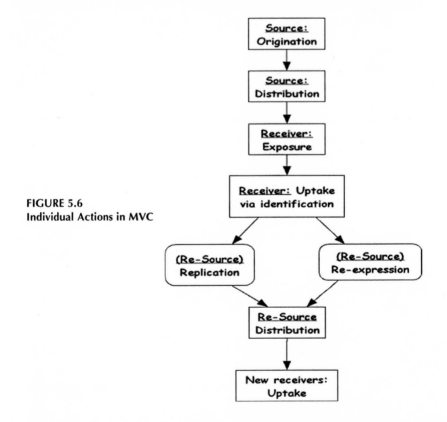

FIGURE 5.6
Individual Actions in MVC

average of twice per day, visiting 105 websites per month, and viewing 2,437 Web pages.[23] Heavy Internet users are also the heavy users of television[24] and radio,[25] so exposure can also occur via traditional media.

Uptake: In medical parlance, uptake means to absorb or incorporate a substance into a living organism. What causes people to take in some content? One broad answer might be identification, a psychological concept that describes cognitive and emotional connections individuals make with other people and content, such as images, songs, performers, and jokes and other texts.

Jenkins refers to these kind of identification processes as part and parcel of the "affective economics" that are characteristic of Internet social media. He describes this effect in a discussion of the viral campaign of LonelyGirl: "They counted on the 'affective economics' of identification others would have for her."[26]

In other words, there is some kind of match between the receiver and the message. Thus, communicators also recognize that characteristics of messages also play a role in the number of users who take it up. This chapter will consider them in a later section.

Replication or Representation: Replication occurs when they pass along the message exactly as they received it. Representation occurs when they approximate the message or actively remix or re-create it, according to their own interpretation or desire.

The reasons people pass along a message are many and varied. They may simply want social interaction,[27] a desire for positive reinforcement,[28] self-expression[29] or self-actualization.[30] Finally, there are limits on passing along content that includes copyright restrictions and the costs of using the Internet, such as access and hosting.[31]

Distribution: Shuen identifies three types of social media users: connectors, brokers, and salesmen. *Connectors* maintain many online relationships and put people together. The other two types are the most likely to send information to others. *Brokers*, or as Shuen calls them *mavens*, offer information. What they don't know, they will find out and tell other users how to find the information they seek. *Salesmen* also provide information, attempting to persuade others to act in a desired manner—buy, vote, volunteer, sign up, recommend, etc.

People are exposed to content whether or not they are connected to the Internet. However, distribution requires resources. In addition to a computer and Internet access, it also requires knowledge of the structure of content dissemination and the computer skills to execute the distribution. These requirements will pose no barriers to professional communicators but they are impediments to a substantial number of people everywhere in the world.

Figure 5.7 shows how the independent uptake and pass-along of many individuals result in MVC, even if it all begins with a message sent by a single individual. The next section examines two models of the overall MVC process as it extends across a population. The first model is based on the theory of the diffusion of innovation.[32] The Shah et al. model[33] is an empirically developed structural model.

Shuen's model begins with the Everett Rogers[34] theory of the diffusion of innovation, which explains how innovations spread throughout a population, including both individual adoption and population adoption. Adopters are people who take up an innovation that could be pottery, personal computers, or messages. Information about the innovation spreads from one individual to another, via communication with interpersonal connections of the adopter. Figure 5.8 shows stages of the individual adoption:

Figure 5.9 depicts the spread of the adoption over time (the gray S-curve) and the distribution of types of adopters (the black normal curve), showing how adopters fall into different categories, depending on when they decide to adopt the innovation.

The Shuen model describes viral communication in terms of a process of adoption that takes place within an Internet network structure. She traces the movement of messages from first-adopters (or first-receivers) and first-degree clusters of receivers through hubs and connectors until there is a small world network of related receivers. To predict the eventual total adoption (or spread of an MVC message), she suggests using a variation of the Rogers S-curve called the Bass Diffusion Curve.[35]

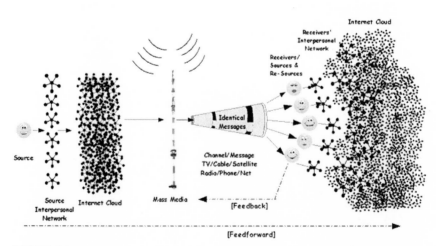

FIGURE 5.7
Process of MVC, Initiated by a Message Generated by an Individual

Five Stages in the Decision Innovation Process

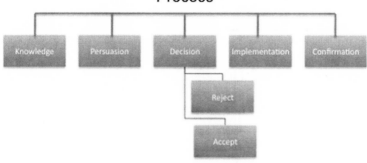

FIGURE 5.8
Stages of the Individual Adoption of Innovation Process

Essentially, this measure lets communicators estimate the likelihood of a message going viral. The Bass Diffusion Curve (BDC) considered diffusion in interpersonal networks, stating that the likelihood of widespread diffusion depends on the number of people who independently adopt something (in this case a message) and the number of people that they influence to also adopt. Plotting the curve requires some calculation, but having a measure of "virality" that allows communicators to monitor the effectiveness of their

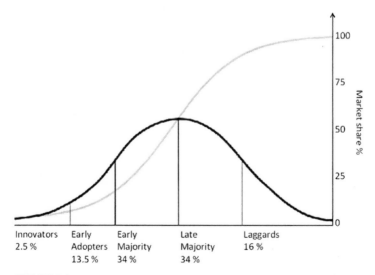

FIGURE 5.9
Stages of Adoption Within a Population

MVC strategy is an important arrow in the campaign quiver. And by focusing on the rate of growth of independent first-time adopters—receivers who first got the message from a mass media source—the BDC also neatly incorporates the influence of mass media.

What would this model predict if there were many blog and discussion group posts, user recommendations and glowing reviews, and other social media mentions? The BDC includes the role of social influence, although the measures might need to be updated. Interestingly, professional communicators now identify "influencers," people who communicate about their interests to many social media contacts. They believe that these influencers will stimulate the number of first-time adopters.[36]

The BDC is a variant of the power law, sometimes called Pareto's Law, or the 80-20 Rule. It is not really a law; it is a rule of thumb, an estimate of many types of distribution, including wealth, Internet use, and a host of other phenomenon. The 80-20 Rule suggests that 20 percent of social media users will have 80 percent of the connections and make 80 percent of the posts.[37]

The 80-20 Rule was supported in one empirical study.[38] The researchers analyzed more than 400,000 blog posts over 30 days to learn how information is diffused by bloggers. In addition to finding that about 20 percent of bloggers were indeed responsible for 80 percent of the blog posts, the study also found that there are three types of topic structure, based on spikes and chatter, as shown in figure 5.10:

- Just spikes
- Spikes + chatter
- Mostly chatter

Gruhl et al. describe the structure of blog traffic. As topics come up on a blog, most of them get attention from a few bloggers and then disappear—the topic is confined to mostly chatter. A few topics gain immediate prominence, often a subject of the mass media. Many people have an opinion and, having expressed it, stop blogging about it—the topic is categorized as just a spike. A very few topics inspire lasting interest, which the authors define as *resonance*. They are also likely to be discussed in the mass media and result in multiple spikes and more or less continuous chatter. Finally, the researchers observed that some authors write many posts on the same subject, and confirmed the presence of influencers, bloggers whose words carry weight, and connectors, highly connected individuals.

The Shah et al. model[39] is an empirically developed structural model that shows how online messaging extends the traditional model of mediated communication. Shah et al. characterize MVC as a series of interdependent activi-

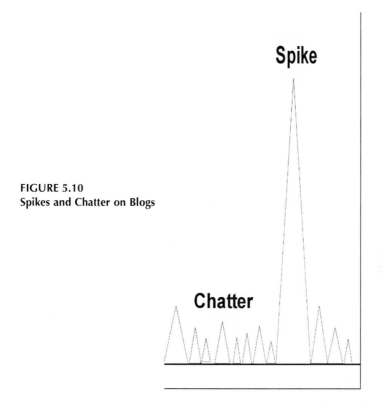

FIGURE 5.10
Spikes and Chatter on Blogs

ties: consuming, participating, and producing. The researchers theorized that the combined exposure of a person to the mass media and Internet use, plus information seeking, political discussion, and interactive political messaging, would all lead to greater political participation, as shown in figure 5.11. (Look at the dotted lines in the theorized model and note that some media exposure might suppress further information seeking.)

This is an unusually helpful study because it looks specifically at online expressive behavior inherent in MVC. Online expressive behavior was characterized as using e-mail to organize, contact politicians or editors, and discuss politics or news, expressing political opinions online, and participating in a chat room or online forum. The researchers measured each of these variables with data from two sources: Campaign Media Analysis Group (CMAG) data from the 2004 election, made available by the University of Wisconsin Advertising Project (WAP) and national panel data, gathered by Synovate for the *Life Style Study* for the DDB-Chicago annual mail survey. Shah et al. measured each of the variables in the model. They tested the pathways all

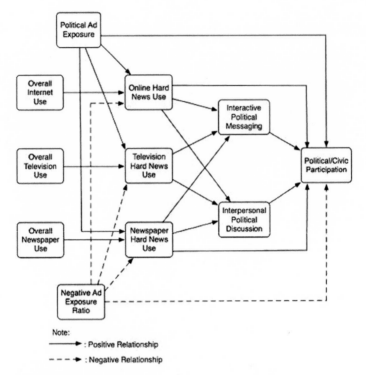

FIGURE 5.11
Theorized Model of Campaign Communication Mediation

together and the measured model supported the theorized one: Mass media exposure, news exposure, interactive political messaging, and interpersonal political discussion did all lead to higher political participation, as shown in figure 5.12. Note that the suppression effect from some media exposure held true as well.

For purposes of looking at MVC, Shah et al.'s most important finding is: "Online news use and interactive political messaging—uses of the web as a resource *and* a forum—both strongly influence civic and political participation, often more so than traditional print and broadcast media and face-to-face communication." In other words, Web use increased participative behavior more than any other communicative activity, whether it occurred via mass media or the interpersonal network. The researchers attribute this increased participation to "expression effects," brought about by the reflection on content and mental elaboration required to re-express previously received content. Thus, engaging in MVC affects the behavior of participants.

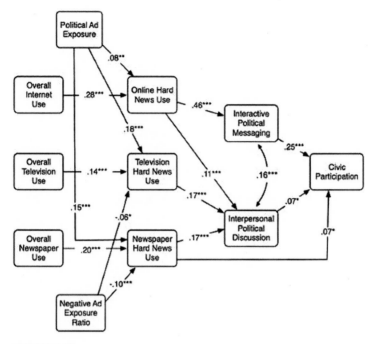

FIGURE 5.12
Influence of Online News Use and Messaging Activity on Civic and Political Participation

Catchy Content

Why do some messages go viral? There is much speculation about this question, but no one really knows for certain.[40] The study by Gruhl et al. showed that most topics that generate online buzz come from the Real World, as opposed to simply arising out of blogosphere chatter. They also observed that topics do not usually capture attention for more than a few days. At the same time, some subjects are clearly more interesting than others, so only these few topics achieve ongoing popularity for discussion.

Grimes theorized that the uniqueness of each meme might stimulate an individual to redistribute it to others: "One can see that the combination of semantic reaction, identification, etc., and the newer, almost novel, memetic mechanisms could offer quite a powerful arsenal for understanding our relationship to each other through communication and our functioning in the extensional world."[41] Heylighen[42] put forward an extensive list of qualities that might increase the likelihood that receivers would pass forward a meme. Such characteristics included utility, novelty, simplicity, coherence, authority and proselytism.

Research suggests some specific dimensions of content that encourage people to consume it include: humor;[43] memorability;[44] and sex and violence.[45] The decision to pass a message along to others generally occurs when a receiver finds the content interesting, important, entertaining, or informative. As a general observation, they pass along material that is attractive and communicable.[46]

In the last three years, there have been many examples of viral communication that reached large numbers of receivers. Some of them suggest these and other characteristics, as presented in table 5.2.

Other scholars have pointed to successful viral messages as being memetic, replicable, and memorable. However, less has been said about the characteristics of it being personal, direct, incomplete, and imperfect. Except for directness, these last four characteristics contrast sharply with most messages, especially those disseminated with persuasive intent. In general, content fashioned by professionals is polished and produced to the highest degree of perfection possible. Even imperfection is carefully calibrated for a precise effect.

Many instances of viral content do not exhibit such care. Rather, they are often expressive, emotional, and haphazardly produced. Yet, these very characteristics may provide an opening for receivers to pass content along to

TABLE 5.2
Characteristics of Messages That Enhance the Likelihood They Will Be Passed Along

Message Characteristic	Description
Memetic	Has cultural resonance or stickiness with which large numbers of people can identify. The mass media play a key role because, the mere appearance of a topic in the mass media can make it a meme.
Replicable	Capable of being copied or re-expressed so that a recognizable version of the meme is transmitted through some combination of creativity, imitation, and duplication
Memorable	Has a hook or quality that people who are exposed can remember. This often means that the content has shock value, such as humor, sex, or violence
Personal	Conveys the sense that the communication has importance and meaning to the sender. Often packs an emotional punch.
Direct	Gives the receiver the feeling that the message is directed at him or her. Often employs some form of direct address such as "you" or "we."
Incomplete	Leaves room for the person to add to the message
Imperfect	Does not intimidate receivers by setting production or execution standards they cannot meet when they re-source the content

others, particularly as material for re-expression than mere forwarding. By allowing the receiver to enter the content, to see ways to add to it creatively or intellectually or otherwise to make it their own, sometimes figures large in MVC messages. As Hoechsmann and Cucinelli note: "Enabled by an economy of viral, point-to-point, communication, where media messages flow on horizontal axes from producers to consumers, some YouTube producers have found mass audiences for the expression primarily of point of view narratives."[47]

Harnessing MVC

Some MVC efforts seem random and accidental—Chris Crocker's *Leave Her Alone* video crying about the press hounding Britney Spears,[48] the many video spoofs of the Saturday Night Live skit, *Lazy Sunday/Chronicles of Narnia* (original archived on http://www.hulu.com), and many jokes and chain e-mails that appear in almost everyone's inbox.

However, many (and perhaps most) MVC events are purposive campaigns, which have been conceptualized, produced, and implemented by professional communicators: the Hotmail viral e-mail tag, subservient chicken, Lonelygirl15, the viral videos that supported the Obama candidacy for president in 2008, Yes, We Can and "lost by one vote" video, and others. Although most of these examples are consumer marketing efforts, the Obama campaign's use of MVC underlines the case that there is no inherent reason MVC will not be effective in fulfilling a range of communication objectives. This section considers what managers need to understand about mounting a MVC effort.

In each of the cited campaigns, there is a creative idea that links both the dissemination of the message and for the message itself. This dual creative challenge is a new aspect of communication plans. The *media mix* (the selection of channels used to transmit the message to a target audience) has always been an essential element of campaign development. However, it was developed largely independently of the creative effort. In MVC plans, the media mix is tightly linked to the message, as well as to the consumers' media habits, which are assumed to include some form of Internet communication. As Dobele et al. note: "Successful viral marketing campaigns are comprised of an engaging message that involves imagination, fun and intrigue, encourages ease of use and visibility, targets credible sources, and leverages combinations of technology."[49]

Traditional managers may find the lack of control over feedforward campaigns such as MVC troubling. There have been negative results from viral efforts, such as viral marketing for *Blair Witch Project*, Sony's portable

PlayStation handheld device, and Campbell Soup. In each case, lack of authenticity caused consumer backlash.[50] Carl[51] found that when marketers disclosed their corporate connections, receivers trusted their information more than when they did not disclose.

A few principles emerge from the literature on VMC: (1) listen to users (consumers, audiences, supporters, etc.); (2) create opportunities for users to interact with one another, not with the company; (3) information is still power—but it's not the information you keep, it's the information you share; (4) identify and target influencers; and, (5) give users something to do, someone to contact, and someplace to go.

What does the future hold for MVC efforts? A 2008 survey[52] queried 40 executives at U.S. advertising agencies. The report found that marketing professionals know about VMC and interest levels are high. Indeed, they intended to increase their VMC budgets in the coming year. They consider the primary benefits of VMC campaigns to be exponential reach and high levels of consumer engagement. At the same time, while they are generally optimistic about the potential for success of VMC campaigns, they are also concerned that metrics for tracking and reporting campaigns are insufficient and the criteria for defining success are still unclear.

Media Managers Going Viral

The skills for executing VMC campaigns have changed substantially over the last few years. Creative thinking and messaging have always been essential. Today, managers need a formidable mix of creative and analytical skills.

The data dredging and analysis involved in identifying, understanding, and reaching users are vital parts of every contemporary communication campaign. Understanding research methods and data analysis is now essential for almost everyone involved in the campaign, not just for the number-friendly MBA executive. Even before looking at the results, practitioners must assess research itself: Does it ask the right questions, does the data collected provide answers to those questions, does the data conform to previous experience and to the observations of people working in the trenches, does it provide insights?

VMC campaigns are just that—campaigns. They are rarely one-off efforts. There is a sequence of messages that must be created as a communication arc. However, there must also be built-in flexibility to accommodate in-process tweaks to campaign elements that are not working effectively. Thus, managers must also be able to evaluate in-progress campaign evaluation research.

Will it provide them with regular data as the campaign progresses about how many people are reached, how many pass along messages, and how many act in ways encouraged by the campaign (click, buy, attend, vote, support, volunteer, donate, etc.)?

The science is key—but so is the art. Once valid, reliable data is in hand, then creative messaging can target the right people at the right time with the right message. The ability to enter another person's motivations and desires is an art, but today, it is married to science. Messages are evaluated by how closely they tap into the targeted users' thoughts, feelings, and lifestyles. But producing that message requires the leap of creativity.

Because creativity plays such an important role in VMC, managers need to know how to manage creative workers. Creatives differ from traditional blue-collar and white-collar employees,[53] requiring more autonomy and self-actualization. Another key difference is that creative workers may have greater expertise than managers, so that the typical superior-subordinate relationship, where the manager knows more than the supervised, is turned on its head.

MVC Next

The technologies to carry out MVC campaigns will continue to evolve. Bandwidth is growing as cable and telephone companies recognize that their customers increasingly use the Internet platform to access visual media. The demand for bandwidth is particularly strong for personal mobile media, such as cell phones. New high-capacity 4G networks will roll out over the next five years and wireless wideband networks will become more ubiquitous throughout public spaces.

The development of Internet 2, now largely confined to government agencies, the military, and academic institutions, will deliver torrents of data to users. It is likely to take as long as a decade to create the software to manage a large number of users who receive and send enormous files. The build-out of physical networks is likely to take even longer.

As bandwidth grows, devices shrink. The science to produce nano transceivers to receive and send streams of data (content) is developing rapidly. The Dick Tracy watch communicator is coming. And perhaps even implanted transceivers will be possible in the future.

MVC is part and parcel of the many-to-many communication modality enabled by the Internet. Social networking technologies are becoming more and more linked, creating a federation of interconnected platforms. Aggregating services such as www.ping.fm, www.meebo.com, and social news, bookmarking, and recommender sites like www.digg.com, www.reddit.com, and www.epinions.com are bringing about densely connected services and users that are making MVC occur in faster cycles that reach large numbers of global users.

It is communicators' opportunity and dilemma: Good news travels fast—so does bad news.

Notes

1. Sean B. Carroll, "Evolution at Two Levels: On Genes and Form," *PLoS Biology* 3, no. 7 (2005): e245.

2. Michael Noll, *The Evolution of Media* (Lanham, MD: Rowman & Littlefield Publishers, 2007).

3. Ronald S. Burt, *Contagion: Models of Imitation and Interpersonal Influence* (2004), 1–32, http://faculty.chicagobooth.edu/ronald.burt/teaching/Contagion.pdf (28 November 2009).

4. Douglas Rushkoff, *Media Virus!: Hidden Agendas in Popular Cultures* (New York: Ballantine Books, 1994).

5. Richard Dawkins, *The Selfish Gene* (New York: Oxford University Press, 1976).

6. Robert G. Grimes, "General Semantics and Memetics: A Tentative Relationship?," *Et Cetera* 55, no. 1 (1998): 30–33.

7. Dawkins, 195.

8. Bruce Edmonds, "The Revealed Poverty of the Gene-Meme Analogy—Why Memetics per se has Failed to Produce Substantive Results," *Journal of Memetics—Evolutionary Models of Information Transmission* 9, no. 1 (2002), http://cfpm.org/jom-emit/2005/vol9/edmonds_b.html (18 November 2009).

9. Scott Atran, "The Trouble with Memes: Inference Versus Imitation in Cultural Creation," *Human Nature* 12, no. 4 (2006): 351–381.

10. Matt Gers, "The Case for Memes," *Biological Theory*, 3, 4 (Fall 2008): 305–315.

11. Wilbur L. Schramm, *The Process and Effects of Mass Communications* (Urbana: University of Illinois Press, 1954).

12. Elihu Katz, "The Two-Step Flow of Communication: An Up-to-Date Report on an Hypothesis," *Public Opinion Quarterly* 21, no. 1 (Spring 1957): 61–78.

13. Kathleen K. Reardon and Everett M. Rogers, "Interpersonal Versus Mass Communication: A False Dichotomy," *Human Communication Research* 15, no. 2 (1988): 284–303.

14. Fahri Karakas, "Welcome to World 2.0: The New Digital Ecosystem," *Journal of Business Strategy* 30, no. 4 (2009): 23–30.

15. "Definition Of: Connectivity," *PC Mag.com*, (n.d.) http://www.pcmag.com/encyclopedia_term/0,2542,t=connectivity&i=40241,00.asp/ (16 November 2009).

16. Carl Bialik, "Marriage-Maker Claims Are Tied in Knots," *Wall Street Journal*, July 29, 2009, http://online.wsj.com/article/SB124879877347487253.html (14 November 2009).

17. Nicole B. Ellison, Charles Steinfeld, and Cliff Lampe, "The Benefits of Face-book 'Friends': Social Capital and College Students' Use of Online Social Network Sites," *Journal of Computer-Mediated Communication* 12, no. 4 (2004): 1143–1168.

18. Mark S. Granovetter, "The Strength of Weak Ties," *The American Journal of Sociology* 78, no. 6 (May 1973).

19. Robert D. Putnam, *Bowling Alone: The Collapse and Revival of American Community* (New York: Simon & Schuster, Inc., 2000).

20. J. Donath and D. Boyd, "Public Displays of Connection," *BT Technology Journal* 22, no. 4 (2004): 71–82.

21. Sydney Jones and Susannah Fox, *Generations Online in 2009* (2009), Pew Internet & American Life Project, http://www.pewinternet.org/~/media//Files/Reports/2009/PIP_Generations_2009.pdf (4 December 2009).

22. Francis Heylighen, "Objective, Subjective and Intersubjective Selectors of Knowledge," *Evolution and Cognition* 3, no. 1 (1997): 63–67.

23. Francis Heylighen, "Average Internet User Visits 100 Sites Per Month," *Rto Online*, April 14, 2008, 2008, http://www.rtoonline.com/content/Article/Apr08/AverageInternetUsage041408.asp/ (28 November 2009).

24. Wayne Friedman, "Double Play: Big Web Users are Big TV Viewers," *Media Daily News*, October 31, 2008, http://www.mediapost.com/publications/?fa=Articles.showArticle&art_aid=93900 (28 November 2009).

25. An Nguyen and Mark Western, "The Complementary Relationship Between the Internet and Traditional Mass Media: The Case of Online News and Information," *Information Research* 11, no. 3 (April 2006), http://informationr.net/ir/11-3/paper259.html (28 November 2009).

26. Henry Jenkins, *Convergence Culture* (New York: University Press, 2006).

27. Katelyn Y. A. McKenna, Amie S. Green, and Marci E. J. Gleason, "Relationship Formation on the Internet: What is the Big Attraction?," *Journal of Social Issues* 58, no. 1 (2002): 9-31.

28. Elizabeth Joyce and Robert E. Kraut, "Predicting Continued Participation in Newsgroups," *Journal of Computer Mediated Communication* 11, no. 3 (2006): 723-747.

29. Kaye D. Trammell and A. Keshelashvili, "Examining New Influencers: A Self-Presentation Study of A-list Blogs," *Journalism and Mass Communication Quarterly* 82, no. 4 (2005): 968-982.

30. Guosong Shao, "Understanding the Appeal of User-Generated Media from the Uses and Gratifications Perspective," *Internet Research* 19, no. 1 (2009): 702–722.

31. Julie L. Russo, "User-Penetrated Content: Fan Video in the Age of Convergence," *Cinema Journal* 48, no. 4 (Summer 2009): 125–130.

32. Amy Shuen, *Web 2.0: A Strategy Guide* (Sebastopol, CA: O'Reilly Media, 2008).

33. Dhavan V. Shah, Jaeho Cho, Seungahn Nah, Melissa R. Gotlieb, Hyunseo Hwang, Nam-Jin Lee, Rosanne M. Scholl, and Douglas M. McLeod, "Campaign Ads, Online Messaging, and Participating: Extending the Communication Mediation Model," *Journal of Communication* 57 (2007): 676–703.

34. Everett M. Rogers, *Diffusion of Innovations* (New York: Free Press, 1962).

35. Frank Bass, "A New Growth Model Product for Consumer Durables," *Management Science* 15, no. 5 (2969): 215–227.

36. Amy Shuen, *Web 2.0: A Strategy Guide* (Sebastopol, CA: O'Reilly Media, 2008).

37. Adam L. Penenberg, *Viral Loop* (New York: Hyperion, 2009).

38. D. Gruhl, David Liben-Nowell, R. Guha, and A. Tompkins, "Information Diffusion in Blogspace," in *WWW2004 held in New York, May 17–22, 2004* (ACM, 2004): 491–501.

39. Dhavan V. Shah, Jaeho Cho, Seungahn Nah, Melissa R. Gotlieb, Hyunseo Hwang, Nam-Jin Lee, Rosanne M. Scholl, and Douglas M. McLeod, "Campaign Ads, Online Messaging, and Participating: Extending the Communication Mediation Model," *Journal of Communication* 57 (2007): 676–703.

40. Richard E. Caves, *Switching Channels* (Cambridge, MA: Harvard University Press, 2005).

41. Robert G. Grimes, "General Semantics and Memetics: A Tentative Relationship?," *Et Cetera* 55, no. 1 (1998): 30–33.

42. Francis Heylighen, "Objective, Subjective and Intersubjective Selectors of Knowledge," *Evolution and Cognition* 3, no. 1 (1997): 63–67.

43. Tim Padgett, "What's Next After that Odd Chicken?," *Time*, October 8, 2004.

44. Vanda Carson, "Uproar Over Nokia Ad Video," Snopes.com, 2003, http://www.snopes.com/photos/commercials/nokia.asp/ (18 March 2008).

45. Lance Porter and Guy J. Golan, "From Subservient Chickens to Brawny Men: Comparison of Viral Advertising to Television Advertising," *Journal of Interactive Advertising* 6, no. 2 (Spring): 30–38.

46. Mark Schaller, Lucian Gideon Conway, III, and Tracy L. Tanchuck, "Selective Pressures on the Once and Future Contents of Ethnic Stereotypes: Effects of the Communicability of Traits," *Journal of Personality and Social Psychology* 82, no. 6 (2002): 861–877.

47. Michael Hoechsmann and Giuliana Cucinella, "My Name is Sacha: Fiction and Fact in a New Media Era," *Taboo* (Spring-Summer 2007): 91–111.

48. Video found at http://www.youtube.com/watch?v=kHmvkRoEowc&feature=related.

49. Angela Dobele, David Toleman, and Michael Beverland, "Controlled Infection! Spreading the Brand Message Through Viral Marketing," *Business Horizons* 48, no. 2 (March-April 2005): 143–149.

50. Sean Carton, "Five Rules of Viral Marketing," *ClickZ*, February 5, 2007, http://www.clickz.com/3624847/ (8 December 2009).

51. Walter J. Carl, "The Role of Disclosure in Organized Word-of-Mouth Marketing Programs," *Journal of Marketing Communications* 14, no. 3 (2008): 241–245.

52. *Viral Video Marketing Survey: The Agency Perspective* (Los Angeles, CA: Feed Company), Feed Company, http://www.feedcompany.com/wp-content/uploads/Feed_Company_Viral_Video_Marketing_Survey.pdf (2 December 2009).

53. John Hartley, *Creative Industries* (Oxford, UK: Blackwell Publishing, 2005).

II

IMPLICATIONS OF NEW MEDIA TECHNOLOGIES

6

The First Domino

The Recorded Music Industry and New Technology

Robert Bellamy and Robert Gross

THE NEWS IS FRIGHTENING FOR TRADITIONAL MEDIA. Broadcast television continues to lose viewers and advertising revenues to cable and satellite television. The Comcast purchase of NBC is an example of just how much the fortunes of the broadcast networks have changed in the last few years.[1] The problems of newspapers have become so prominent they are now plot points in popular entertainment. In the last couple of years, such major papers as the *Rocky Mountain News* and *Seattle Post-Intelligencer* have ceased publication and others (*Detroit Free Press, Washington Times*) have deliberately slashed circulation and delivery in an attempt to survive.[2] The recent demise of such once popular and profitable magazines as *Gourmet, Teen,* and *Vibe* reveals that the problems in print media are hardly limited to newspapers.[3] More ominously, each of these magazines were specialized publications. The lesson is that not even content aimed at a specific niche audience is guaranteed success in the present media environment.

While the problems of broadcast and print media have gained the most attention in terms of recent popular coverage, the recorded music industry (RMI) was the first to feel the effects of a new age of media driven by digitalization and the Internet. If, as appears increasingly evident, we are entering a new age of media characterized by more consumer choice and control, at least for those who can afford it, the canary in the coal mine was the record business.

The first decade of the 2000s has brought nothing but bad news to the RMI. Saying that the industry is in a desperate down spiral would not be hyperbolic. Consider that recorded music sales have plunged by about one-third in the decade.[4] Revenues have declined from a peak of $14.6 billion in 1999 to $10.8 billion in 2008.[5]

Consider also that:

1. More and more major artists (e.g., AC/DC, Eagles) are making their own deals with big-box distributors such as Walmart and Target at the expense of both record labels and traditional music sales outlets.[6]
2. Some major artists, with Radiohead being the most prominent example, have bypassed both the big boxes and big labels in offering music directly to the public.[7]
3. Such major artists as Robbie Williams, Madonna, and Jay-Z have pioneered "360 deals" whereby all facets of their economic power from recorded music sales to concerts to merchandising are all now part of a single artist contract. These deals are interesting and important because some of them do not involve traditional record companies.[8]
4. Economic consolidation has led to an oligopoly of only four major worldwide recorded music companies (Universal Music Group, Sony BMG, Warner Music Group, and the EMI Group) with further mergers likely.[9]

Consolidation has also swept the live music business with Live Nation now controlling the vast majority of regular concert venues in the U.S.[10] Live Nation is in the process of merging with Ticketmaster, the near monopoly seller of entertainment tickets in the U.S.[11] The combined revenues of these two companies (over $12 billion) will dwarf that of the entire RMI.[12] As for radio, the various formats that are labeled "Rock," long the most popular format when its various sub-genres are aggregated, has seen a steady decrease in listenership, as radio also deals with reduced advertising revenue.[13] This is a symptom of a changing RMI, as more and more consumers and fans prefer the program-your-own "formats" they can create on their MP3 players or obtain from such Internet services as Pandora or Slacker. Since radio and recorded music have had a symbiotic relationship for well over fifty years, the radio industry, or at least the terrestrial broadcast side of the industry, is also suffering as a result of the radical recent changes in how people obtain and consume music.

On the retail side, the big boxes of Walmart, Target, and Best Buy, themselves some of the few survivors of consolidation within the discount department and electronic/appliance store industries, have now become dominant in the physical sale of recorded music. Tower Records, Camelot, NRM: National Record Mart, Sam Goody, Media Play, and Circuit City, among many other once major sellers, are no more. More tellingly, the largest recorded music retailer in the nation is now the iTunes Store.[14] Those who buy physical

CDs are now the minority of music purchasers.[15] Most troubling of all, a considerable amount of downloading is done without payment. While iTunes, Amazon, Rhapsody, et al. have made legal downloading easy and relatively inexpensive, an entire generation of music consumer has grown up with illegal downloads normalized within the culture. Anyone who has asked a class of undergraduates about their music acquisition behavior can verify this assertion. Beginning with Napster over ten years ago and then with Grokster, Lime Wire, Bit Torrents, Gnutella, eDonkey, Morpheus, and other methods, illegal downloading has had a tremendous impact on the traditional music business. Although there is some evidence that illegal downloading might be easing,[16] there have been estimates in the last few years that the rate of piracy for music might be as high as 95%.[17]

Downloading is particularly commonplace among the young, a group that has long been dominant in the consumer base for recorded music. This explains why recorded music is the first, but definitely not the last, major media industry to be forever altered by the new media. It also represents a very troubling trend for the traditional powers in the RMI. As the young age their habits may modify slightly, but are unlikely to radically change. Music downloading is essentially here to stay. The best the industry can hope for is a lessening of illegal downloading in favor of the legal. The days of the general interest and well-stocked record store is nearing its end. The era of a few powerful recorded music companies acting as a cartel to control the production, distribution, and sale of recorded music may not be at an end, and is doing everything it possibly can to avoid that fate, but it has changed beyond all recognition in slightly over a decade. This might even be an optimistic assertion. Steve Knopper, industry analyst and author, claims in his recent book *Appetite for Self-Destruction*, that as of the end of 2008, "it sure feels like the end is near" for the traditional RMI.[18]

The purpose of this chapter is to explain how and why things have changed so rapidly in the recorded music industry. In addition, we will offer perspective on what the future of the recorded music industry is likely to look (and sound) like. Historical context is necessary to understanding the present predicament of the RMI. In particular, a history of the relationships the record industry has had with other media and with its customers will help demonstrate that the current crash has antecedents beyond the changes in technology of the last 10-20 years.

A Brief and Selective History of the Recorded Music Business

The senior Oliver Wendell Holmes wrote that "[music] is to the soul what water is to the body."[19] Music can invoke powerful memories while simultaneously eliciting a vast array of emotions unique to and dependent upon an individual's own experience. These properties, among others, make music perhaps the most abstract of all media. Music, like painting and sculpture, has for centuries been considered an art form. Of all the media, perhaps only books and movies have such artistic cachet. But, more than any other media form, music has the almost mystical powers of individualized impact. Music is something that people can do themselves. Most people will never write a book or produce movies, but most of us have participated in some form of music making whether it be banging out teenage angst with three chords and a guitar, being a member of a choir or chorus, playing an instrument in a school or garage band, or simply singing off-key in the shower or in the car. More than any other medium, music belongs to the average person.

Because of the intimate connection of music and everyday experience, the music business began its commercial rise by both presenting popular entertainers in touring stage shows and by selling sheet music of popular songs so people could make their own music. The first time in human history that anyone could hear a sound, at least one that others could also hear, coming from the ether without a wire and not directly from another person, animal, or object, was the first time they heard a "record" player. From the cylinders of the Edison era to the many innovations of Berliner, Siemans, and many others, a record player did not take long to become a must-have home appliance, at least in the more prosperous homes in the country. Recording technology also allowed the rustic "people's music" originally labeled as "race" (blues, R&B, jazz) and "hillbilly" (country, C&W) to reach a new, growing, and eventually nationwide audience. By the mid-twentieth century the "Tin Pan Alley" and Broadway tunes of the urban areas of the U.S., and particularly New York, were joined by many other forms of both indigenous and borrowed music from the rural areas of the nation where the bulk of the population lived at that time.[20]

Of course, the commercialization and professionalization of radio in the mid- and late 1920s was critical to these developments. Radio allowed people all over the nation to hear both local and national talent. While the more "raw" (or as some have said, more "authentic") amateur voices were eventually all but banished from the nation's airwaves, a considerable number of people were able to make a good living as radio and recording artists.

Radio was an almost perfect medium for the transmission of music to a mass audience because of its audio-only nature and the fact that advertiser-supported broadcast radio was "free" to the consumer. Of course, the music received by the early radio listeners was live, because recording technology with reasonable quality was not available until after World War II, and few stations had any incentive to play records when there was so much network and local live entertainment available. As tape became prominent after the war, the broadcasting industry essentially won a battle with the American Federation of Musicians, which broke the ban on and the stigma of playing recorded material on radio.[21]

The Music/Radio Symbiosis

While the end of the dispute made recorded programming increasingly common on radio, other technological changes were to completely change the nature of the radio and music industries. In 1949, Columbia introduced Peter Goldmark's 33 1/3 rpm Long Play (LP) recording format.[22] The LP made it possible for one record to have numerous songs or space for longer pieces, i.e., as common in symphonic and some jazz music. The record-buying public would no longer have to constantly change the old and relatively brittle 78 rpm records that had been the industry standard. The LP, of course, was necessary to the development of the concept of the "album" as a coherent single piece of artistic expression, a concept that is declining in the download age.

Paralleling the 33 1/3 LP was RCA Victor's development of the 45 rpm format. These small and inexpensive "singles" were to become the medium for the explosion in the popularity of recorded music among the young.[23] In addition, new and less expensive record players, eventually to include Hi-Fi (high fidelity) and then stereo as standard, were allowing more and more households to acquire record players. Indeed, having a "hi fi" or console stereo was a clear sign of the middle-class "good life" in the postwar explosion of consumerism. Of course, another sign of the good life was the acquisition of a television set.

The advent of television changed all existing media. General-interest magazines (*Life, Look, Saturday Evening Post, Collier's,* etc.) eventually folded as they could not compete with a new medium that could bring both sound and vision into the comforts of the household. The motion picture business had to completely change its oligopolistic business structure because of the advent of television and legal decisions (*United States v. Paramount Pictures,* 1948).[24] The not-so-old studio system began to fade away and the product of the industry became more specialized with particular appeal to the burgeoning youth market.

The youth market was also the salvation of both the record industry and the radio business as television became, as it remains today, the most dominant home medium. The RMI began to actively promote a relatively diverse group of artists to a growing youth market. Included were many artists that once would have been relegated to such sub-genres as race and hillbilly. Most important, the mixture of R&B, country, and Tin Pan Alley resulted in the new genre of "Rock 'n' Roll" by the mid-1950s.[25] While the beginnings of the rock revolution inspired many budding musicians and led even more to go see the music performed live, most of it was consumed on records purchased in stores *and* played on the radio.

The enormous popularity and rapid expansion of television took an enormous toll on the traditional role of the radio industry, as the new medium took away much of radio's programming and a great deal of its audience. Network radio, once the lifeblood of most major radio stations, withered away as a programming source throughout the 1950s, leaving behind news, sports, and a few talk/service programs. The soap operas, children's programs, and, especially, prime-time programs all faded away as television became the primary home entertainment medium. Radio operators were forced to create their own programs and attract new audiences to survive.

The beginning of the Rock era in music almost exactly parallels these changes in radio. The new music was seen by many savvy radio operators, both network and non-affiliated stations, as their salvation. Instead of the expense of producing or acquiring programs, radio used one announcer, a collection of singles, and blanket licenses from the ASCAP and BMI music licensing organizations to make the Disc Jockey (DJ) and "Top 40" formats the iconic radio images from that time forward.[26] This switch to virtually all recorded music formats in radio was heavily supported by the RMI once it realized how sales of records and sales of tickets for appearances by the artists were spurred by radio airplay. The innovation of portable and transistor radios and the increasing ubiquity of car radios also spurred the development of music radio, particularly among the fast-growing baby boomer group of young people.

The symbiotic relationship of the RMI to radio was further fueled by a "Wild West" mentality that eventually was at least partially criminalized. Many record companies and promoters were more than willing to compensate poorly paid and barely supervised DJs for playing certain records. Payola was a welcome source of revenue for station personnel and, has been documented many times in historical and more popular accounts of the era, a key factor in "breaking" any number of new musical acts to a wider audience.[27] Clearly, such a system can only exist when there are both willing buyers and sellers of a desirable project. The haphazard nature of this first period of payola is an example of an evolving industrial relationship; i.e., one with many smaller entities attempting to break through to mainstream success. Many radio stations, so long dependent on network programming, were attempting to completely change their mode of operation in television's wake. The RMI, with its relatively low barriers to entry for entrepreneurs, attracted an entirely new type of businessperson . . . one who "hustled" and was willing to take chances in terms of artists signed, the breaching of racial and socioeconomic barriers, and any number of business practices. Whether involved in payola or not, such labels as Sun, Chess, Imperial, Philles, Dot, Hi, and dozens of others were able to carve out a space in a burgeoning new industry.

The symbiotic relationship of the recorded music industry and the radio industry was well established by the late 1950s, and both industries entered a long period of prosperity and popularity. Radio and records were *the* media of teens and young adults by the 1960s, while television primarily was the medium of young children, older adults, and families. The popularity of both media to the young was fueled by the various shades and sub-genres of Rock that became increasingly popular throughout the 1960s and beyond.

Stagnation and Recovery I

Although the period from approximately the mid-1950s to the late 1990s were generally prosperous ones for the record business, the very excitement and success of that industry led to periods of retrenchment, change, and consolidation. For example, the success of early Rock 'n' Roll triggered the well documented backlash against many of the early rock stars because of their alleged bad behavior and failure to conform to social norms and, just as important from an industrial standpoint, the early promoters and producers of rock music. The results of this backlash were the cracking down on payola; the hounding and legal troubles of such early notables as Jerry Lee Lewis, Little Richard, Chuck Berry, and Alan Freed; and the attempt of major companies tied into the larger entertainment business (ABC/Paramount, CBS/Columbia, RCA/NBC) to stake out much larger shares of the popular music business. These large corporations did this with much safer (i.e., clean-cut, white) and mediagenic artists as the "teen idols" (Frankie Avalon, Paul Anka, Pat Boone, et al.) who could also become popular television and motion picture personalities.[28]

The received wisdom is that this early period of stagnation ended with the "British Invasion" of 1963-64 led by the Beatles and soon many other bands and artists (The Rolling Stones, The Who, Donovan, Herman's Hermits, et al.) who, somewhat ironically, gained great success by their homages to many of the R&B musicians so influential in the formation of Rock 'n' Roll or, paradoxically, by their appeal to the folk revival or teen idol fads of the early 1960s. While this received wisdom is incomplete, ignoring, for example, the rise of Bob Dylan and others in the emerging "folk-rock" genre, the Beach

Boys and their intricate pop sound, the pop/R&B of Motown, and even the "Countrypolitan" sounds coming out of Nashville, there is no doubt that the British invasion was a major part of the rapid changes taking place in the RMI in the 1960s and 1970s.[29]

Another key event in the RMI's revival was the increasing recognition of the album as a coherent music statement rather than just a couple of singles and filler. Rock criticism in the pages of *Crawdaddy*, *Creem*, and *Rolling Stone* encouraged serious consideration of popular music as an important component of the arts in more and more mainstream publications. The fact that many artists were now songwriters as well as musicians was another major change as the power of songwriters and songwriting "factories" (e.g., Brill Building) went into steady decline as power moved inexorably to artists, their managers, and producers. While the "Top 40" formats popularized in the 1950s by Gordon McLendon and others remained popular for many years to come, the radio industry tracked the changes in the recorded music by introducing new sub-genres of Rock such as AOR (album oriented rock), Urban (i.e., R&B and soul aimed primarily at black audiences), etc.[30]

The popularization of FM transmission, with its superior sound and multiplexing possibilities (i.e., stereo radio) was another major boon to both music and radio's popularity in this time period. The superior sound of FM and the consequent growth in the sales of better radio receivers and home and car stereos, with first eight-track tape and later cassette players, all spurred the record industry to new heights of popularity and profits.

Stagnation and Recovery II

By the mid- and late 1970s, the record industry was again in a slump due to its own business practices and demographic changes in society. The aging of the baby boomer generation negatively affected the sale of albums and singles during this time. Artists who at one time had been a beneficiary of youth appeal (e.g., Rod Stewart, Paul McCartney, and country artists such as Charlie Rich and Kenny Rogers) became bland MOR (middle-of-the-road) artists. These moves were successful in the short run, but further alienated the core customers of recorded music. There may not have been as many young people as a few years early, but they still bought music far out of proportion to their numbers. Disco music, while wildly popular in clubs, was a poor genre for sales (with some major exceptions) and, to many observers, seemed to spell an end to the serious AOR of previous years.[31]

Another factor in this period of decline was radio programming. Music selection had been essentially taken out of the hand of idiosyncratic, hard-to-control, and even corrupt DJs and shifted to Program Directors (PDs). As part of management, PDs were much too risk averse to "break" music that had not already been vetted by trade industry charts and the growing profession of music consultants who devised formats that sounded greatly alike from market to market. The homogenized sound of radio was a serious turn-off to many listeners and potential record buyers.[32]

The RMI was increasingly vilified for these changes. By the end of the 1970s, six major record labels (EMI, Sony, BMG, PolyGram, WEA and MCA) dominated domestic and international record sales. Many of the once innovative independent record companies had gone out of business or were acquired by one of the Big 6. While more competitive than other oligopolistic industries of the day (the broadcast television networks, the domestic auto industry), the six majors followed the typically oligopolistic pattern of a lack of innovation (i.e., even disco music originated with smaller companies) and what many regard as price gouging. Even during the recessionary years of the late 1970s and early 1980s, the majors continued to raise the price of their product to an audience that was less and less responsive to its wares. Major retail chains, often located in shopping malls, were the places where most people purchased music and these retailers had little incentive or were even prohibited from reducing the retail price of recorded music when there was so little competition for sales. For music consumers, it was definitely a time of "take it or leave it" with many choosing the latter.[33]

The 1980s recovery of the record industry was fueled by three major factors. First, the rise of rap/hip-hop and the many sub-genres derived from this music was a major spur to record sales. New artists, producers, and record companies (Def Jam, Sugar Hill, Death Row, etc.) created an excitement in the industry not seen since the 1950s. While the commodification of rap was to come, the new music, initially disdained by white-controlled businesses and many radio stations, became part of the emerging American zeitgeist of combinations of existing forms into something new . . . whether in music, sexuality, gender, race, or other cultural touchstones.[34]

Second, MTV was a huge boon to the recorded music industry as television, for the first time, became popular among the youth who contribute so mightily to record sales. While actual viewership of MTV was relatively limited, the new channel had a power much greater than such viewership would suggest. Within just a few years, the idea of music as a visual as well as an aural medium was well established. MTV and the artists it favored received enormous attention from other media. Even the negative attention the network received (i.e., "corroding" young minds; sexist, racist, and misogynistic content) was a spur to sales just as such criticisms of early Rock 'n' Roll had made that music more desirable by the young in a previous generation.

Sales of traditional albums and singles, as well as music videos, soared in the 1980s due in part to the power of MTV. MTV not only inspired other cable music channels (CMT, Ted Turner's Cable Music Channel, and VH-1, its own spin-off network), it also legitimized music as important to the entire television and motion picture industries. By the early 1990s, for example, the placement of music in programs or movies was a growing source of promotion and revenue for the music business. MTV, of course, was also in the first wave of niche cable programming services that demonstrated that a combination of cable subscriber fees and advertising (i.e., the dual revenue stream) could generate enormous sums of money and make cable competitive with broadcast television.[35]

The third major factor in the record business's 1980s revival was the innovation of the compact disc (CD). CDs became enormously popular within months after their commercial introduction in the U.S. in 1983.[36] Anyone who visited a record store in the mid-1980s was likely to see CDs on the counter and out of their sleeves to emphasize their durability and small size. Although many disagreed, CDs were also advertised as providing better (i.e., digital) sound quality than LPs. Whether or not the quality of a CD was superior to that of an LP was not an issue that weighed heavily, if at all, in the minds of most consumers.

As has been demonstrated time and again in the history of media technology, an innovation with demonstrably lesser quality (e.g., VHS, RCA color television service) is often the winner in the marketplace. The reasons for this range from the market power of manufacturers to political decision making to simply an audience preference for the convenience of a new medium at the expense of quality. In short, audiophiles complained about the quality of CDs in the 1980s while the mass of consumers swarmed to this new method of distributing music. In fact, a major impetus for the growth of CD sales was the replacement market, where people who were older than traditional record buyers purchased CD copies of their old albums. The creation of this replacement market was recognized by *Billboard* and other analysts of the industry with the creation of new charts that tracked "catalog" sales.[37]

The Comeuppance

The major increases in revenues and profits for the recorded music industry in the 1980s and 1990s were not sustainable. This is to be expected of an industry that has always operated within up-and-down cycles of success. These cycles are particularly pronounced because of the particular resonance of demographic and cultural changes on the music business. The difference in the next (and current) slump was that there seemed to be no way for the industry to extricate itself and reach another upswing in the cycle.

Digitalization that had been the key to the conversion to the CD format became a serious problem when digital files became widely available on the Internet. Television increasingly became a competitor rather than an enabler for the industry as its power to break artists and spur sales became increasingly important to recording artists. Video games, ringtones, Internet radio, and other websites (MySpace, YouTube, etc.) were all to become new ways in which the music consumer could find what she or he desired. In addition, many artists in the burgeoning rap and alternative rock genres found ever increasing ways of circumventing the power of the industry. Obviously, the reasons for the recent downfall of the RMI are numerous and complex, but can be summarized principally as a failure of the industry to understand its primary customer, a lack of understanding that had the effect of further alienating that customer.

The fact is that there has never been a particularly good relationship between the consumers of music and the record industry. While price gouging and a failure to release desirable music are important components of this attitude, there are others. Since the beginning of serious music criticism in the 1960s and, in some cases even back to the early payola scandals, the RMI has been portrayed in a negative way. This portrayal eventually spread to the mainstream media and became part of the conventional wisdom. Record companies were regarded as being run primarily by people who knew or cared little for music or the artists who made it. The companies were part of giant out-of-touch corporations that "screwed" both artists and the public.[38] With a few exceptions (e.g., , the Ertegun brothers at Atlantic, David Geffen in his early career with Asylum, Clive Davis in his later career with Arista and J Records, and several of the independent rock and rap producers such as Russell Simmons and Rick Rubin), people in the record industry were treated much more negatively in the press than even people in other media industries. These negative perceptions rarely applied to recording artists who, no matter how much money they made, were seen as being at the mercy of rapacious and oligopolistic record companies. Artists such as the Beatles, Rolling Stones, Prince, Madonna, and Pearl Jam who established (or were granted) their own "independent" record companies or had long-running battles with their labels were considered to be heroes because they supposedly won their battles against the big labels. In the late 1970s, artists like Tom Petty and the Heartbreakers gained great credibility by protesting the raising of retail prices, a move paralleled by Pearl Jam's battle against Ticketmaster in the 1990s.[39] In other words, the record industry for all its success had few defenders among its customer base or even among most recording artists.

This alienation became even more pronounced in the late 1980s and 1990s as it became evident that the major record companies continued to raise the cost of music to the consumer while their own costs (i.e., the cost of physical production of a compact disc is lower than that of a vinyl record) dropped. In addition, in a repeat of historic industry practice, a new wave of pre-teen-oriented "boy-" and "girl-" bands and artists (e.g., Spice Girls, N Sync, Back-street Boys, Britney Spears) appeared, to the dismay of many slightly older customers. This was combined with a seemingly cavalier attitude toward older artists and new artists that led to another round of content bashing.[40]

At the same time that these primarily singles artists were popular, the industry reduced the number of single recordings available. This meant that fans would have to buy an entire album to get the one or two songs they wanted. This was a particular issue for children and teens who had less disposable income than adults and further alienated them from the industry.[41] Even Country music radio and record labels, which had a long tradition of keeping artists active on the chart for decades, began to discard pioneers of the genre (Johnny Cash, George Jones, Loretta Lynn, et al.) in favor of "hat acts" and pop-flavored artists and songs. However, this time the audience had an alternate to reducing or eliminating their purchases only to reinvigorate them a few years later.

Diffusion and Relative Advantage

One of the lessons that can be learned from a study of the research on diffusion of innovations and, more particularly in terms of media innovations, is that software is by far the most important factor to the consumer. Hardware may be an essential conduit but has minimal intrinsic value to the majority of consumers. Hardware or to be more exact the medium itself is only important insomuch as it provides the content (software) that the consumer desires. This is a rather broad simplification, but an important one. This is because the history of the media industries is replete with failed hardware that appealed to a relatively small portion of "mediaphiles." Landfills are (or were) the final destination of many once heralded technologies (e.g., Betamax, LaserDiscs, AM stereo).

Of course, there are many reasons that a new innovation becomes popular in the marketplace. Cost and accessibility are important components of diffusion. Another key is relative advantage.[42] This is where the consumer/user accepts a new technology as a replacement for an existing one because it is seen as better than the previous technology. Relative advantage leads to a substitution effect, which, not surprisingly, is much greater if there is a cost advantage with the new innovation. Most innovations, of course, both cost more than existing technology and are more difficult to use. Both of these factors tend to impede diffusion. However, in the case of recorded music, the innovation of downloading already had a distinct cost advantage over the traditional way of acquiring music. In addition, the main audience for recorded music had minimal difficulty in using the new technology/distribution mechanism of digital downloading. In fact, these advantages were so pronounced that the recorded music industry was totally unprepared to do anything but lash out at its best customers.

Piracy has long been a concern of and problem for the recorded music industry. Technological innovations in audio recording long before the commercialization of the Internet made it relatively easy to copy material. Piracy is estimated to cost the economy more than the total value of the entire recording industry as well as tens of thousands of jobs and millions of dollars in lost tax revenues.[43] The Recording Industry Association of America (RIAA), the major trade/lobbying group for the RMI has long spent considerable resources on fighting copyright infringement. Before the Internet became ubiquitous, a major issue for the record business was home taping, a concern that grew as tape became a near universal audio accessory that allowed for the creation, trading, and selling of copied music. Tape also allowed music consumers to become their own "programmers" through the creation of mix tapes, and for artists to create and distribute "demo tapes" to fans and potential employers. Of course, both of these reduced the economic power of the record business. These concerns became more pronounced with the introduction of Digital Audio Tape (DAT) in the late 1980s. DAT allowed for the creation of "perfect" digital copies of copyrighted material. The RIAA was unable to stop DAT, but was able to gain relief through the *Audio Home Recording Act* of 1992 that forced the new medium to include anti-circumvention processes to limit copying and to pay fees to the record industry to make up for potential lost revenues.[44] The industry was further protected by the *Digital Millennium Copyright Act* (DMCA) of 1998 which, among other provisions, raised the penalties for violating copyright via the Internet and criminalized the breaking of anti-circumvention codes even if no copyright infringement occurred.[45]

Obviously, copyright protection is an important and valid concern as the protection of intellectual property is one of the cornerstones of the entertainment industry. In a capitalist economy individuals and corporations have the right to profit from their efforts to create intellectual property. While we cannot dispute the legal rights of any industry to protect its property from illegal piracy, the recording industry made some very bad decisions in its effort to combat the problem.

Napster was the key player at the beginning of the change in the fortunes of the recorded music industry upon its introduction in June 1999. Developed primarily by Sean Fanning, Napster allowed for the trading of music files via the Internet.[46] Napster had near immediate appeal for the young. More specifically, Napster had particular resonance with college students as they were more likely than others at that time to have access to relatively powerful computers and fast Internet connections, have the "know-how" to use the service, and to have a strong interest in music. Napster's message was clear—Get the type of music you want free *and* also make a statement to the recording industry that it was no longer meeting your needs. Obviously, the free music angle was the strongest motivator, but the political statement about the industry being made should not be ignored. As discussed, the RMI has a long history of alienating its best customers.[47]

The Industry Responds to Downloading

Although the RIAA had legislative success in the 1990s, earlier judicial decisions on copyright were a potential problem as it attempted to fight downloading. The U.S. Supreme Court had affirmed the right of consumers to copy recorded materials for personal use in the landmark *Sony v. Universal* case of 1982, when the Court ruled that VCRs did not violate copyrights just by their presence in the marketplace.[48] The RIAA adopted the tactic that this was a much different situation because Napster existed almost solely for the violation of legitimate copyrights and seriously undermined record sales. After all, argued the industry, what else was Napster for but the distribution of such copyrights without paying fees? The other problem was that Napster ran its service with centralized servers to connect music "traders." The argument was ultimately successful as the original Napster was found to be in violation of copyright laws in 2001 by the U.S. Court of Appeals (9th Circuit), and consequently closed down.[49]

The RIAA's victory in the Napster case was not as great a victory as the industry might have hoped. Other services that did not use centralized servers but rather allowed users a "hands-off" conduit (software) to share files person-to-person (P2P) such as Grokster, KaZaA, and Morpheus were soon available to everyone that had access to a computer and a relatively fast Internet connection. The RIAA, along with the motion picture industry, filed a copyright infringement suit against Grokster. Grokster's argument that it was not guilty of infringement because it did not store or transfer music was successful at the federal appeals level, but dismissed by the U.S. Supreme Court in

2005. Writing for the majority, Justice Souter said that it was "unmistakable" that the primary intent of Grokster was to facilitate illegal acts.[50]

Unfortunately for the recorded music industry, the proverbial "cat was out of the bag." By the time of the industry's court victories, tens of millions of people had downloaded music files from the Internet, with over 20 million estimated to have done so as early as 2001.[51] Digital downloading behavior was ingrained in vast numbers of primarily young people, many of whom were not willing to give up such activity regardless of what the courts or the industry told them. More and more file sharing services became available, with some based offshore to avoid U.S. law. In essence, Napster, Grokster, and the others had taught an entire generation of recorded music's formerly best customers that stealing was not wrong when it was from a large impersonal record company.[52] The normalization of both legal and illegal downloading was further enhanced by the growing number of artists who encouraged downloading and made their music available on MySpace, their personal websites, and other places on the vast Internet. Using the Internet to acquire music, whether legal or not, became the standard by the early 2000s for an entire generation of people between middle school and college age, a normalization that was unlikely to be reversed as this group aged.

The RIAA's legal actions on file sharing were successful, but their other tactic of aggressively going after those who illegally downloaded was a public relations disaster. Between 2003 and 2008 the RIAA adopted a program of seeking legal action against P2P downloads. By the end of the program, an estimated 30,000 people in the U.S. had faced legal action.[53] The problem was that the publicity for seeking legal action against large pirates was virtually non-existent. Rather, the people who were prosecuted for illegal downloading were almost always college students, minors, or even "average" moms and dads, people who almost always generated a great deal of sympathy. The RIAA increasingly looked and acted like a bully. This image was reinforced by the recording industry's refusal to come up with a convenient and cost friendly way to implement a pay downloading service.

The RIAA attempted to reach an accommodation with Napster in 2000. Several industry observers believe such a deal would have been a win/win for the then Big 5 and the consumer and a way for the industry to quickly adapt to new technology. Although details are relatively scant, no agreement was reached in part because the Big 5 supposedly wanted too much control over the service and wanted to limit the degree to which consumers could use Napster.[54] After Napster was forced to close down in 2001, the industry introduced two new download-for-pay subscription services. Press Play was backed by Sony, Universal, and EMI while Music Net was backed by EMI,

Warner, and BMG. The fact that these services did not have music from all the Big 5, and only a limited amount of independent music, was a major problem. The individual corporations in the Big 5 were in essence at war among themselves and with their potential customers. In addition to their limited offerings, both of these services had severe limits on the number of copies and downloads a customer could have as well as strict prohibitions on storage of downloaded tracks on hard drives. An audience that was not accustomed to such "hassles" rejected both services.[55] Former RIAA President Hilary Rosen believes that these failures were when the industry "lost the users. That's when we went from music having real value in people's minds, to music having no economic value, just emotional value."[56]

Despite Rosen's view, a new way of having people pay for music was on the horizon. Apple's iTunes Store, online in 2003, was a major breakthrough in the history of the RMI—one that the industry was not particularly happy with but had no choice but to support. Apple is a company noted for innovation and style, and one that has developed a virtual cult following among a proportion of the population and among the press. Apple introduced its first iPod digital music player in 2001. The advantages of this device over portable cassette recorders and CD players in terms of size and capacity were immediately obvious.[57] The iPod is one of the most transformative media technologies of the first decade of the twenty-first century because it is a key factor in making downloading music from the Internet a normal activity of an ever increasing number of people of all ages.

The iTunes Store, of course, is the major engine for the success of the iPod by offering an ever increasing amount and diversity of music and video for sale for relatively low prices on an à la carte basis. While the record industry was fighting its consumers and, just as important, dithering with developing its own music download services, such companies as Apple, Amazon, and Yahoo! have become distributors of choice for millions of people all over the world.

Parallels with the Past

While it's easy to criticize the RIAA and its members for poor decision making and failure to adapt, criticism that obviously is merited, one should keep in mind we can only point out the RMI's mistakes in retrospect. After all, as the first media industry to be forever changed by new media technology, mistakes were to be expected. After all, the record business was setting sales and revenue records in the 1990s and, no doubt, the good times seemed as if they would go on forever.

Additionally, other media industries have also had bad experiences with technological adaptation. Perhaps the best example is the failure of the motion picture industry to better cope with the introduction of television in the late 1940s. Rather than seeing the production of television programming as an extension of the box office, the movie industry did all it could to hamstring television's early programming by refusing to produce programming and, most disastrously, failing to invest in the new medium as station and network owners.[58]

The motion picture industry failed to understand the substitution effect and the relative advantage that the public saw in and received from television. For those who had the money to buy a television set, the new medium was the replacement for going to the movies. Combine this with the rise of suburbanization, which prevented easy access to the downtown-based movie theaters and other entertainment and it is easy in retrospect to see why television was so popular so soon.

The failure of the movie industry to grasp the importance of the television industry was combined with the existing relationships the radio industry had with the federal government as licensees. This relationship made it relatively easy for the radio stations to apply for and receive most of the early television licenses. After all, it was all broadcasting, a federally licensed "public interest" industry.

As for programming, the early television providers at both the local and national levels transferred programs from radio and supplemented them with relatively inexpensive but visually interesting programming such as sports, cooking programs, and demonstrations of various types. The refusal of the motion picture industry to cooperate had no measurable effect on the popularity of the new medium. By the mid-1950s, of course, most of the movie studios were program suppliers to the broadcast television networks as the motion picture industry adapted to the new media environment of the day.[59] Although the movies' power and influence on American life has never again been as powerful as it once was, the industry continues to prosper and set box office records.[60]

The traditional power and influence of the recorded music industry has similarly lost its once-strong hold on a large segment of the population. Technological change and diffusion are obviously the most important factors. However, the failure of the industry to understand that their role in getting music to listeners was always fragile because of the arrogance and bad publicity with which the RMI has long suffered. There was little public or policy loyalty to the industry when the apocalypse came.

Unless one owns stock in or works for a major record company, there is no particular reason to mourn the present situation in the industry. These are still large international corporations that have further consolidated into the now Big 4 of Universal, Sony BMG, Warner, and EMI. While their traditional oligopolistic power over what we listen to has been breached and devalued, they remain the primary power in the industry in terms of the commercial packaging of artists. They are still the major components of the star-making apparatus in music. The Big 4 continue to generate enormous amounts of money and dominate the worldwide music business, controlling approximately 80% of the U.S. and 70% of the world market for recorded music, a market worth approximately $30 billion.[61]

A Cloudy Future

The problem for the Big 4 is that this packaging is no longer a near monopoly. Major artists can and have completely bypassed traditional record companies to sign "360" deals. On the retail level, stores devoted exclusively to new releases are rapidly disappearing and even the big box powerhouses like Walmart, Best Buy, and Target are reducing shelf space for recorded music and are demanding and receiving more exclusives and special promotional considerations from the industry. Apple continues to dominate the music download industry, with the Big 4 having apparently failed in their collective and individual efforts to create their own digital distribution channels.

The RMI needs to create new and exploit existing revenue streams if it is ever to pull itself out of its ongoing reversal of fortune. Recent efforts to force radio stations to pay for the music they play, rather than getting virtual carte blanche to play what they wish under the ASCAP and BMI licenses, are a potential source of new revenue.[62] Of course, broadcasters vigorously oppose this and have a considerably more powerful political lobby. However, the recording industry has several supporters in the Obama administration that might counterbalance the power of the broadcasters.[63] Such a move could also sever the symbiotic relationship of music and radio with detrimental consequences for both parties. Of course, the recent bankruptcy of Citadel Broadcasting and the continuing economic problems in the radio industry make it clear that the long and fruitful relationship might need to be rethought.[64]

The record business will also have to spend at least as much effort on encouraging legal digital downloading as it does fighting illegal downloading. The RIAA's so-called "educational" program of taking legal action against downloading is now over and should never be repeated.[65] The industry will have to spend considerable resources on teaching people of all ages that downloading is a legal, convenient, and cost-effective way of acquiring music. If it fails to do this, the industry is likely to continue its decline.

The RMI will have to realize that a large proportion of music acquisition will remain free to consumers whether by legal or illegal means. While the advocates of ending copyrights for digital files such as the Electronic Frontier Foundation and other analysts and observers will find little support among the Big 4 or most independent labels,[66] the industry does need to do a better job of adapting to the new reality by working more closely with advertisers to offer free music. This includes more revenue sharing with outlets that use music as a major component of their business, whether that is YouTube or Pandora or *Guitar Hero* video games or ringtones or even concert T-shirt sales. The point is that there are now nearly "endless access points for music" and the industry will have to be involved with as many of these as possible.[67]

Industry observer Gerd Leonhard has written that "music [is] like water" and the only thing the industry can do is "monetize the existing behavior of the user" by developing a way of metering music use and charging for it like water or other utilities.[68] While this seems to be a long-range plan at best, there is sentiment for there to be surcharges on cable and Internet bills to compensate the industry.[69] Whether there is political support for these ideas in time is an open question but knowing how slow the wheels of government often turn and how tax-averse Americans have become, the prognosis for either scheme is problematic at best.

The Big 4 are likely to become a Big 3 or even a Big 2 in the near future as so many of its traditional business functions, particularly in terms of distribution, have been replaced. This assumes that the Antitrust Division of the U.S. Justice Department remains as somnolent as it has been for the last thirty years. The future of the Big 4 (or the Big 3 or even Big 2 of the near future) as separate corporate entities is shaky at best with mergers or absorption by other entertainment industry giants a likely future. For example, a merger of Live Nation (already merged with Ticketmaster) with one of the current Big 4 would create a potentially powerhouse horizontally integrated entertainment company. The acquisition of a Big 4 member as an adjunct of an already existing media company (shades of the once joint ownership of CBS and Columbia Records) such as News Corp. or Time Warner would give the parent company access to the contracts and catalog of a considerable amount of valuable recordings that they now have to license.

Anderson's long-tail hypothesis says that the media future is a multitude of digital products being perpetually available even if they sale only a few units per tracking period.[70] Since digital storage has little cost, a company can prosper with low sales over a long period of time and also by charging more for specialized offerings. Because of the potential of making profits with the long tail, the continued decrease in the overall value of record companies likely will contribute to such mergers and acquisitions. The trend of divestiture of record companies by corporate parents such as Time Warner's spin-off/sale of the Warner Music Group could easily be reversed.

We need to be careful not to overemphasize the shift of power from industry to consumer that seems to be a major byproduct of the changes in the recorded music industry. There is little doubt that any power accruing to media consumers will be countered by changes at the industrial level. The changes in the RMI over the last decade are as much or more about a shuffling of corporate power as a democratizing of music or more emphasis on "grassroots" music production, although there has been more of both.

The bottom line is there will always be a need for professionals backed by large sums of money to package artists, oversee and produce recordings, publicize and market those recordings, and release (distribute) them to consumers through whatever means available. There will also always be a need for professional musicians to have some level of protection for their intellectual property and having assistance in reaching their public. In short, there will always be a robust market for recorded music produced on an industrial scale. However, that scale has been permanently altered.

The example of the motion picture industry of the 1950s and 1960s is instructive. That industry eventually abandoned the studio system with its thousands of employees and high fixed costs in favor of becoming packagers of motion picture and television products. The companies that remain in the RMI will have to similarly become "consumer marketing companies" to survive.[71] There is no longer a viable long-term market for a recorded music industry that is solely about the production and distribution of recorded music.

The way that music gets to consumers has forever been changed as has the industrial structure that for so long almost completely dominated the business. The radio/music symbiosis is being challenged. Digital downloads, both legal and otherwise, are here to stay, as are the myriad other ways we can obtain music. Recent technological change related to digitalization and the Internet has led to obvious advantages to music consumers and the music artist who can learn to exploit the new world of the recorded music industry.

Notes

1. Tim Arango, "G.E. Makes it Official: NBC will go to Comcast," *New York Times*, 3 December 2009, http://www.nytimes.com/2009/12/04/business/media/04nbc.html (21 January 2010).

2. David Kaplan, "End of the Line for Rocky Mountain News; Misses 150th Birthday by Two Months," *paidContent.org*, 26 February 2009, http://paidcontent .org/article/419-end-of-the-line-for-rocky-mountain-news/ (21 January 2010); Dan

Richmond and Andrea James, "Seattle P-I to Publish Last Edition Tuesday," seattlepi, 17 March 2009, http://www.seattlepi.com/business/403793_piclosure17.html (21 January 2010); Neely Tucker, "Sports Staff Fears End of Wahington Times jobs is near," *Washington Post*, 30 December 2009, http://www.washingtonpost.com/wp-dyn/content/article/2009/12/29/AR2009122902804.html?hpid=moreheadlines (21 January 2010).

3. Greg Bensinger, "U.S. Magazine Advertising Sales Drop Widened in 2009 (Update 2)," *Bloomberg Business Week*, 12 January 2010, http://www.businessweek.com/news/2010-01-12/u-s-magazine-advertising-decline-accelerated-to-18-in-2009.html (21 January 2010); Mercedes Bunz, "Ad-Figures for 2009 Show Not Even U.S. Weeklies are Recession-Proof," guardian.co.uk, 13 January 2010, http://www.guardian.co.uk/media/pda/2010/jan/13/pda-us-weekly-magazines-advertising (21 January 2010); Bill Mickey and Vanessa Votolina, "Vibe Magazine Closes," *Folio*, 30 June 2009, http://www.foliomag.com/2009/vibe-magazine-closes (21 January 2010).

4. Mark Mulligan, "The Decade that Music Forgot (A Brief Glance Back on the 10 Years that Unraveled the Music Industry)," *The Forrester Blog for Consumer Product Strategy Professionals*, 30 December 2009, http://blogs.forrester.com/consumer_product_strategy/2009/12/the-decade-that-music-forgot-a-brief-glance-back-on-the-10-years-that-unraveled-the-music-industry-.html (21 January 2010).

5. Mulligan, 2009.

6. "AC/DC Plan Huge World Tour After Album Release," *Rolling Stone*, 21 June 2008, http://www.rollingstone.com/rockdaily/index.php/2008/07/29/acdc-plan-huge-world-tour-following-album-release/ (21 January 2010); Brian Mansfield, "Exclusives Aim to Pull Music Fans into Stores," *USA Today*, 1 March 2007, http://www.usatoday.com/life/music/news/2007-02-28-exclusive-music_x.htm (21 January 2010).

7. Jaron Lanier, "Is the Internet Turning Creative People into Digital Peasants?," *SonicState.com*, 13 January 2010, http://www.sonicstate.com/news/2010/01/13/is-the-internet-turning-creative-people-into-digital-peasants/ (21 January 2010).

8. Robert Sandall, "Off the Record," *Prospect*, 1 August 2007, http://www.prospectmagazine.co.us/2007/08/offtherecord (21 January 2010).

9. Brian Hiatt and Evan Serpick, "Music Biz Laments 'Worst Year Ever'," *Rolling Stone*, 13 January 2006, http://www.rollingstone.com/news/story/9147118/music_biz_laments_worst_year_ever/ (21 January 2010).

10. Glenn Peoples, "Business Matters: Live Nation, AEG, Online 'Spectators' Vs. 'Conversationalists'," *Billboard*.biz, 21 January 2010, http://www.billboard.biz/bbbiz/content_display/industry/e3idb0553264d55f0013aecc18c6d3c65d2 (21 January 2010).

11. Eliot Van Buskirk, "Live Nation and Ticketmaster Prepare to Merge," *Wired*, 4 February 2009, http://www.wired.com/epicenter/2009/02/live-nation-and/ (21 January 2010).

12. Peoples, 2010; Van Buskirk, 2009.

13. "Radio Industry Revenues Expected to Remain Low in 2009; While Mid and Smaller Markets Hold Up Better," *BIA Advisory Services, LLC*, http://www.bia.com/pr090325-radiorevs.asp (21 January 2010).

14. Brian Hiatt and Evan Serpick, "The Record Industry's Decline," *Rolling Stone*, 28 June 2007, http://www.rollingstone.com/news/story/15137581,the_record_industrys_decline/print (21 January 2010).

15. Phil Gallo, "A Decade of Upheaval for the Record Business," *Live Daily*, 30 December 2009, http://www.livedaily.com/news/a-decade-of-upheaval-for-the-record-business-21078.html (21 January 2010).

16. "IFPI Publishes Digital Music Report 2010," *IFPI (International Federation of the Phonograph Industry*, 21 January 2010, http://www.ifpi.org/content/section_resources/dmr2010.html (21 January 2010).

17. "IFPI Publishes Digital Music Report 2010," 2010.

18. Steve Knopper, *Appetite for Self-Destruction: The Spectacular Crash of the Record Industry in the Digital Age* (New York: Free Press, 2009), 252.

19. "Quotations About Music," *The Quote Garden*, http://www.quotegarden.com/music/html (21 January 2010).

20. Michael Chanan, *Repeated Takes: A Short History of Recording and its Effects on Music* (London: Verso, 1997); Evan Eisenberg, *The Recording Angel: Music, Records and Culture from Aristotle to Zappa*, 2nd ed. (New Haven, CT: Yale University Press, 2005); Christopher H. Sterling and John Michael Kittross, *Stay Tuned: A History of American Broadcasting*, 3rd. ed. (Mahwah, NJ: Erlbaum, 2003); Nick Tosches, *Unsung Heroes of Rock 'n' Roll: The Birth of Rock in the Wild Years before Elvis* (New York: DeCapo, 1999).

21. Sterling and Kittross, 2003.

22. Chanan, 1997; Eisenberg, 2005.

23. Glenn C. Altschuler, *All Shook Up: How Rock 'n' Roll Changed America* (New York: Oxford University Press, USA, 2003); Chanan, 1997; Eisenberg, 2005.

24. *Unied States v. Paramount Pictures, Inc.*, 334 U.S. 131 (1948).

25. Tosches, 1999.

26. Sterling and Kittross, 2003.

27. Kerry Segrave, *Payola in the Music Industry: A History, 1890-1991* (Jefferson, NC: McFarland, 1994).

28. Altschuler, 2003; Charlie Gillett, *The Sound of the City: The Rise of Rock and Roll* (New York: DeCapo, 1996).

29. Altschuler, 2003; Gillett, 1996.

30. "Gordon McLendon," Radio Hall of Fame, http://www.radiohof.org/pioneer/gordonmclendon.html (21 January 2010).

31. Altshuler, 2003; Gillett, 1996.

32. Joe D'Angelo, "Don't Blame File-Sharing for Slumping CD Sales, Study Says," *MTV*, 30 March 2004; http://www.mtv.com/news/articles/1486060/20040330/story.jhtml (21 January 2010); Kris Millett, "Radiohead & the Fall of the Recording Industry," *(Cult)ure Magazine*, 31 October 2007, http://www.culturemagazine.ca/music/radiohead_a_the_fall_of_the_recording_industry.html (21 January 2010).

33. Chris Woods, "The Piracy Genie is Out: Here's How to Catch Him," *World Focus: Fighting IP Theft*, May 2005, http://www.kilpatrickstockton.com/publications/downloads/ThreatResponse.pdf (21 January 2010).

34. The Center for Information & Research on Civic Learning and Engagement (CIRCLE) at Tufts University has a wealth of information on youth attitudes. See

http://www.civicyouth.org/ (21 January 2010); Jeff Chang and D.J. Kool Herc, *Can't Stop Won't Stop: A History of the Hip-Hop Generation* (New York: Picador, 2005).

35. Tom McGrath, *MTV: The Making of a Revolution* (New York: Running Press, 1996).

36. "Compact Disc Hits 25th Birthday," *BBC News*, 17 August 2007, http://news .bbc.co.uk/2/hi/technology/6950845.stm (21 January 2010).

37. Jed Gottleib, "Billboard Charts Old Territory, Adds Catalog Albums to Top 22 List," *BostonHerald.com*, 13 January 2010, http://www.bostonherald.com/entertain ment/music/general/view/20100113billboard_charts_old_territory_adds_catalog_ albums_to_top_200_list/srvc=edge&position=also (21 January 2010).

38. Altshuler, 2003; Gillett, 1996.

39. Elizabeth Blair, "Pearl Jam: Still a Touring Force," *National Public Radio (NPR)*, 19 June 2003, http://www.npr.org/templates/story/story.php?storyId=1301462 (21 January 2010); Robert Palmer, "Petty Wins Battle to Lower Record Prices," *Montreal Gazette*, 14 May 1981, 27, http://news.google.com/newspapers?id=EIsxAAAAIBAJ&sj id=v6QFAAAAIBAJ&pg=1021,1400981&hl=en (21 January 2010).

40. "The A.B.C.: Anti Boy Bands Club," *Mucisians Network*, n.d., http://www .musiciansnetwork.com/network/Anti_Music/The_A_B_C_Anti_Boy_Bands_Club-info58585.html (21 January 2010).

41. Knopper, 2009.

42. Everett M. Rogers, *Diffusion of Innovations*, 5th ed. (New York: Fress Press, 2003).

43. "Piracy: Online and on the Street," *Recording Industry Association of America*, n.d., http://www.riaa.com/physicalpiracy.php (21 January 2010).

44. *Audio Home Recording Act*, 106 Stat. 4737 (1992).

45. *Digital Millenium Copyright Act*, 112 Stat. 2860 (1998).

46. Milam Aiken, Mahesh Vanjani, Baishali Ray, and Jeanette Martin, "College Student Internet Use," *Campus-Wide Information Services* 20, no. 5 (2003): 182-85; Rafael Rob and Joel Waldfogel, "Piracy on the High C's: Sales Displacement, and Social Welfare in a Sample of College Students," *The Journal of Law and Economics* 49 (April 2006), 29-62.

47. Knopper, 2009.

48. *Sony Corp. of Amer. V. Universal City Studios, Inc.*, 464 U.S.417 (1984).

49. *A&M Records, Inc. v. Napster, Inc.*, 239 F.3d 1004 (9th Cir., 2001).

50. *MGM Studios, Inc. v. Grokster, Ltd.*, 545 U.S. 913 (2005).

51. Kembrew McLeod, "MP3s are Killing Home Taping: The Rise of Internet Distribution and Its Challenge to the Major Label Music Monopoly," *Popular Music ad Society*, 28 (October 2005), 521–31.

52. Christian Imhorst, "The 'Lost Generation' of the Music Industry," *Datenteiler*, 17 November 2004, http://www.datenteiler.de/translations/the-lost-generation-of-the-music-industry/ (21 January 2010); Knopper, 2009.

53. David Kravets, "Leaner RIAA Still Moving to Terminate Online Access of Copyright Scofflaws," *Wired*, 3 March 2009, http://www.wired.com/threatlevel/ 2009/03/leaner-riaa-sti/ (21 January 2010).

54. Hiatt and Serpick, 2007.

55. Hiatt and Serpick, 2007, Knopper, 2009.

56. Hiatt and Serpick, 2007.

57. Leander Kahney, *The Cult of iPod* (San Francisco: No Starch Press, 2005).

58. Tino T. Balio, ed., *American Film Industry*, rev. ed. (Madison: University of Wisconsin Press, 1985); Barry R. Litman, *The Motion Picture Mega-Industry* (New York: Allyn & Bacon, 1998).

59. Balio, 1985; Litman, 1998.

60. "Analysis: Media Forecast 2010," *Mediaweek*, 3 January 2010, http://www.mediaweek.com/mw/content_display/news/media-agencies-research/e3id839769e9efdcc3a9aafa7d23c15c2ec (21 January 2010).

61. Mulligan, 2009.

62. Brian Wingfield, "Radio Royalties Fight Heats Up in Washington," *Forbes*, 8 January 2010, http://www.forbes.com/2010/01/08/radio-internet-royalties-business-beltway-radio.html (21 January 2010).

63. Declan McCullagh, "Obama picks RIAA's favorite lawyer for a top Justice post," *CNet News*, 6 January 2009, http://news.cnet.com/8301-13578_3-10133425-38.html (21 January 2010).

64. Michael J. de la Merced, "Citadel Broadcasting Files for Bankruptcy," *New York Times*, 20 December 2009, http://www.nytimes.com/2009/12/21/business/media/21citadel.html (21 January 2010).

65. Kravets, 2009.

66. "The Free Music Philosophy (V1.1)," *Electronic Frontier Foundation*, n.d., http://w2.eff.org/IP/Audio/?f=free_music.article.html (21 January 2010).

67. Hiatt and Serpick, 2007; Jeff Howe, "Why the Music Industry Hates *Guitar Hero*," *Wired*, 23 February 2009, http://www.wired.com/culture/culturereviews/magazine/17-03/st_essay (21 January 2010).

68. Gerd Leonhard, "Music Like Water," *New Music Box*, 25 July 2005, http://newmusicbox.org (21 January 2010), http://www.newmusicbox.org/article.nmbx?id=4311 (21 January 2010).

69. David Kusek, "Music Like Water," *Forbes*, 31 January 2005, http://www.forbes.com/forbes/2005/0131/042.html (21 January 2010); Leonhard.

70. Chris Anderson, *The Long Tail: Why the Future of Business is Selling Less of More.* (New York: Hyperion, 2006).

71. Hiatt and Serpick, 2007; Knopper, 2009.

7

Changes and Challenges in the Print Industry

The New Landscape of the Print Media

Steven Phipps

O F ALL THE MAJOR FORMS OF MODERN MASS MEDIA as they are usually defined, the medium of print is the oldest. In spite of a host of innovations that have improved the medium since its inception, print technology has remained surprisingly stable for centuries. Now, new digital methods stand poised to transform the print media into something entirely different, and no one knows for sure what form its ultimate incarnation will take.

Development of the Print Industry

The printing of books was, of course, prefaced by handwritten production of manuscripts. Texts were committed to scrolls of parchment, papyrus, or vellum, as well as inscriptions in marble, impressions in pottery, and jottings on sheets of parchment. During the period of the Roman empire, libraries were established in order to house the collected intelligence of the then-known world in the form of, for the most part, handwritten scrolls.

By the third century BCE, an alternative to the scroll was already in limited use. Someone reasoned that stacking individual sheets atop one another and stitching them together along one edge would result in a greatly improved format. The radical new media technology that resulted became known as the codex, from the Latin *caudex*. Unlike the scroll, the codex facilitated quick and easy access to any particular page. As a result, the world's first random-access media format had been created.

The oldest extant codex has been dated to about 100 CE,[1] but scrolls were used throughout the Roman Empire until a gradual displacement began to occur in the third century. By that time, early Christians were popularizing the codex as their words spread around the Mediterranean. The result was that the form achieved widespread adoption.[2]

Mass communication histories have tended to claim that book production moved directly from scribal production to production by printers with no intervening process. One observer points out,[3] however, that this view omits what she calls "a previous move from scriptoria to stationers' shops." During the twelfth century, approximately three centuries before the introduction of printing in Europe, lay stationers began to replace monastic scribes.

Books were still produced by a handwritten process, but in the context of a factory-like system. For that reason, stationers' shops could be viewed as the first media businesses in the modern sense. No longer were books created only as personal expression. Now, for the first time, they had become mass-produced products created by collective commercial endeavor.

By the fifteenth century, the popularization of printing by Johann Gutenberg in Germany forever changed the complexion of Western society. No longer were the scriptures reserved for an elite few. As the tenets of reformers like Martin Luther were committed to print, the Reformation changed the fabric of European society and culture. Gutenberg's printing process inaugurated what could arguably be termed the first medium truly designed for the masses. As a result, the printed material that soon flooded Europe radically altered the church, science, politics, and the arts.

America's first printed book is generally said to have been a nearly 300-page volume titled *The Whole Book of Psalmes*, commonly referred to as the Bay Psalm Book. This book was published in 1640 in the Massachusetts Bay colony, but its publication did not inaugurate anything like a flourishing publishing industry.

On the whole, printing was slow to develop in colonial America as compared to England. This was primarily because of two factors:[4] First of all, America lacked a sufficiently large population in concentrated areas to support the purchase of expensive books. Second, full-scale publishing of books of substantial caliber required ample capital that would need to be repaid quickly. Book publishing was not an industry that lent itself well to rapid return on investment.

For these reasons, it was generally more cost effective to simply import enough books from England to satisfy the colonies' relatively paltry needs than to encourage indigenous printing. A colonial publisher would have been reluctant to invest in high quality publication of a major book title as long as a local bookstore could easily import a small number of copies of a similar title published in England and gain immediate sales.

By the 1790s this picture had begun to change. This was mainly because of the break with England, but also because a few enterprising and forward-thinking American publishers had begun to take measures to protect investments in publishing ventures.

Benjamin Day, credited with the introduction of the first "penny press" newspaper in 1833, envisioned a newspaper that could support itself from ad sales far more readily than by subscriptions. A low copy price, he reasoned, would support a large readership. A large readership, in turn, would enhance the paper's attractiveness to advertisers and this, in turn, could enable him to increase ad rates.

This fundamental concept is immediately analogous to that of Ted Turner in the mid-1970s. Turner pioneered the use of telecommunication satellites to distribute television programming to cable systems, who would in turn route that programming to their subscribers.

If his nondescript UHF TV station in Atlanta made its signal available to cable systems all over the country by satellite, Turner reasoned, the increased viewership would enhance the value of advertising, and increased ad sales could be used to fund more attractive programming. This would, in turn, further enhance the value of advertising on his fledgling superstation and later cable networks.

In much the same way, Benjamin Day made far more money in the newspaper industry by dramatically decreasing his prices. As the demand for his newspaper rose, he could more readily offer the public just the type of entertainment they wanted. As he did so, circulation would continue to rise, and so would the ad rates.

Benjamin Day forever altered the face of the print media in much the same way as had Ted Turner with satellite-distributed cable TV beginning in the mid-1970s. Those alterations sent ripples through not only the newspaper industry but the entire realm of book publishing as well. For the same reasons that the penny press quickly gained ground as a true mass medium, the book industry began to reorganize along the lines of profitability and mass acceptance.

As the publishing industry, along with advanced distributed networks, began to develop, so did the demand for certain titles and authors. The best seller in the nineteenth century was the Bible, followed by Harriet Beecher Stowe's novel *Uncle Tom's Cabin*. In the latter half of the century, "dime novels" and paperbacks were huge sellers.

As a result, promotion began to take on a life of its own. The book industry was extensively rationalized in the early twentieth century. By this time, the lone individual publisher had all but vanished except in small niche markets. Bank financing, sophisticated promotional tactics, and literary agents all played important roles by this point. The paperback market, which traces its beginnings to the nineteenth century, greatly expanded after World War II. Modern trends in book publishing include mergers (a current driving force in all major media) and online sales.

As a result of the developments outlined above, the requisite economic and cultural factors were in place by the end of the twentieth century to allow for continued growth and development in the print industry. By the end of the first decade of the twenty-first century, however, monumental changes in the media landscape in general were beginning to threaten the status quo, with revolutionary implications for the traditional print industry.

The Shift to Digital and its Potential for the Future

Audio and video media have increasingly been moving toward a new system of production, storage, distribution, and delivery of content. That system is dependent on two essential components: dependence on digital forms for all aspects of both visual and aural media, and use of an interactive content-on-demand system, which in turn facilitates access and payment, where payment is necessary. In the future, essential components of this system can be expected to be applied to what has formerly been known as the print media.

Revolutionary change in the former print media can be expected once three vital processes are in place. First, producers, distributors, and consumers of materials formerly made available only through print must come to recognize the value of digital-based "virtual" books, magazines, and newspapers as viable forms. Second, hardware must become widely available that will make electronic reading more attractive than the reading of paper-based materials. Third, a complex combination of legal and economic issues must be addressed.

Already, electronically distributed digital text-based media are beginning to supplant traditional print as the media of choice in some applications. Newspapers are feeling the pinch, largely because of competition from the Internet. Print newspapers depend primarily on revenue from advertising. Newspaper ad revenues have been sharply down,[5] but what is not clear is to what extent this is due to the economic recession as opposed to Web-based competition.[6]

The viability of print newspaper advertising is, of course, tied to the value of the newspaper as a primary news source. Because the Internet offers instant access to frequently updated news reports, print newspapers' trust in older advertising models is being undermined.[7] Talk of charging for access to content placed online by print-based newspapers is being revived,[8] although newspaper websites would then have to compete with television stations and networks, which typically offer online news for free.

Extensive discussion both in the news media and in Congress have centered on a perceived need for saving what might be a dying newspaper industry.[9] Even where newspapers are still showing a profit, that profit appears to be in decline rather than in growth mode.[10] At the same time, the future for the traditional print-based newspaper is uncertain. While some might see this as a non-issue since new media can be expected to rush in to fill the void, critics of this perspective point out that relatively little original reporting is initiated by websites.[11]

The Google Books initiative may have served to encourage public acceptance of ebooks, simply because of its high visibility and because of the expansive scope (about seven million books by the end of 2009) of the project. Controversies surrounding Google Books have perhaps served as the best example of the sort of convulsions that have been rocking the print industry as the result of the emergence of digital forms.

These controversies resulted in a class action lawsuit filed by publishers and writers, with preliminary approval of a settlement granted by the court late in 2009. A primary concern raised in the lawsuit was Google's dominance over its ebook initiative.

The company has partnered with a number of major libraries in its efforts to scan books and other printed materials. In their haste to participate in the digital learning community, however, libraries may have overlooked the fact that Google's expansive project is, after all, part of a profit-making venture by a single company.

In general, access to books, magazines, and other printed materials included in Google Books varies. Degree of access ranges from barely more than a mention, with links to possible print version suppliers, to complete access in online and even downloadable form. Factors that determine access have to do with whether the book or other item is out of print and whether it is still protected by copyright.

In the short term, the mere presence of Google Books might accomplish much in terms of publicizing and popularizing ebooks and ebook options. In the long run, however, even the availability of public domain titles for free via Google Books might exert some negative impact on the economic viability of the ebook industry. This is because publishers who are dependent, at least in part, on public domain materials could be effectively denied the opportunity to republish them at a profit.

The well-known 2003 Supreme Court case known as *Eldred v. Ashcroft*[12] examined the constitutionality of the Sonny Bono Copyright Term Extension

Act. The case highlighted the fact that some print-based publishers make at least part of their living by reprinting formerly copyrighted works that have entered the public domain.

If all public domain printed materials become freely downloadable through Google Books, one could easily envision such publishers being forced out of business entirely. This could only be expected to happen once the use of handheld ebook readers becomes commonplace.

Challenges and Issues Surrounding the Switch to Digital Forms

Several factors have hindered the widespread adoption of handheld ebook reader devices until recently. Certainly one such factor has been a general unfamiliarity with those hardware options that already exist. The public has tended to confuse reading from a handheld ebook reader with the uncomfortable process of attempting to extensively read text from the screen of a typical desktop computer.

Handheld ebook reader device adoption has also been hindered by dependence on proprietary formats even for public domain titles. Rather than pursuing an open source model, the ebook reader industry has, until very recently, been typically characterized by the lack of a reader that will directly accept commonly used file formats. A vast number of public domain texts are freely available in txt (text) or PDF (Adobe Acrobat "Portable Document") formats through such Web initiatives as Google Books, Internet Archives, and Project Gutenberg.

Handheld readers, however, have generally demanded the use of formats unique to that particular device. In some cases, conversion is possible, but through an e-mail process and for a fee or, alternatively, by using such third-party software as Mobipocket.

The recent introduction of a new Kindle product, the Amazon Kindle DX, might mark a permanent step in the direction of open source ebooks. Unlike the conventional Kindle, the Kindle DX allows for direct input of PDF and text files without the need for conversion. In addition, its larger screen size (two and a half times larger) more readily supports display of unconverted PDF files.

Another encouraging development in the spread of ebook technologies is the recent adoption of the EPUB format as a download option for public domain titles included in Google Books. EPUB, which stands for "electronic publication," is unlike its more popular counterpart, PDF, in that it does a better job of resizing text to fit the small screens of smartphones and other miniature handheld devices. Perhaps widespread adoption of EPUB and PDF for ebook reader direct input could stave off an otherwise inevitable ebook format war.[13]

Use of standard file formats such as PDF and EPUB could also obviate an initial failing of Kindle's pre-DX proprietary format. Publishers complained that the format had problems handling tables and monospaced text, two features that are especially valuable in technical publications.[14] An across-the-board switch from PDF to EPUB would probably not be desirable, because some books that are heavily dependent on special formatting are deemed to require PDF.[15]

The release of popular new and lucrative titles in either EPUB or PDF could be a problem without some form of digital rights management (DRM) technology. If concern about copyright infringement would impel the introduction of yet another format, perhaps such a format could be used only for books that are protected by copyright. Use of non-proprietary formats for public domain texts could encourage the growth of ebooks generally.

In addition, despite much-heralded advances in screen technology, hand-held ebook reader screens could still be far better before they could be said to be comparable to print quality, although that level of quality is already being claimed. Screens offering merely dark gray text against a lighter gray background cannot be expected to compete successfully with print with all consumers.

Substantial impetus toward development of better quality screens has, however, resulted in steady improvement. Electronic paper displays can be nearly as flexible as paper. In addition, they are not dependent on backlighting. Backlit monitors may be great for television and desktop computers, but are hard on the eyes of those who attempt to read lengthy electronic texts.

Electronic paper reflects light much like traditional paper. Experimentation has been in progress for at least the past forty years or so to develop a viable form of electronic paper. Efforts have focused on the use of microparticles which offer high contrast without drawing electrical power.[16] This can result in a surface that somewhat resembles traditional paper, with text and graphic display provided by the responses of microparticles to electrical signals.

These responses depend on a process that is sometimes referred to as "electronic ink," and the ink is embedded in electronic paper. A system developed by Joseph Jacobson at MIT, for example, uses micro-encapsulated pellets—10,000 of them per square centimeter—that appear white when subjected to a negative charge, or black in response to a positive charge.[17] These pellets can be thought of something analogous to pixels, but without the troublesome backlighting of conventional computer displays.

Electronic ink was first used in an ebook reader in 2004, when Sony introduced the handheld Sony Libre. The Libre was not financially successful, however, and was discontinued. Persistent rumors of an Apple tablet-type computer, which appeared in January 2010, had centered around its potential as an ebook reader.

In addition to electronic ink, another technological factor in the success of the Kindle and other recently introduced devices is a minimal use of battery power. This is a byproduct of the use of electronic ink, allowing users to go for days and even weeks without the need for recharging the device.

Out of all the many handheld ebook reader devices that have been introduced, the various forms of the Amazon Kindle have, in general, already revolutionized the concept of ebooks. The Kindle was introduced in 2007, and since that time its already acceptable screen quality has improved.

Kindle users do not need to find a Wi-Fi hotspot in order to download books and periodicals. Instead, materials can be downloaded from almost any location, using the cell phone network. (Technical issues surrounding this latter feature have, however, slowed down introduction of the Kindle in other parts of the world).[18]

The Kindle's significance for the development of the ebook industry lies in the fact that it is the first ebook reader device to gain widespread acceptance and even popularity. Amazon has attempted to tie the purchase of ebooks to its own sales efforts, much like the iTunes store for iPod users. If, however, the ebook industry makes a marked turn in a decidedly more open source direction, then Amazon might be bypassed by consumers in some applications.

Not all publishers are alike in their level of enthusiasm for the Kindle. Some of them find the available file formats too limiting, while others are turned off by the concept of all their ebooks being distributed only through Amazon.[19]

Amazon's domination of the ebook market through the popularity of the Kindle might become overshadowed by the introduction of Apple's iPad, introduced January 27, 2010. The iPad is 0.5 inches thick and weighs 1.5 pounds. Consumers can use the iPad to check and send e-mail, watch videos, listen to music, play games, read books, and many other applications.

Further rumors suggest that Apple is eyeing the tablet as a potential contender for the ebook reader device market. In fact, as one observer put it, Apple "has big plans for ebooks" and is currently engaged in discussions with publishing executives representing a major chunk of the industry.[20]

According to one unconfirmed report, Amazon is currently believed to be offering publishers a 50/50 revenue split. Apple is said to be offering, instead, a 30/70 share in the publishers' favor. According to the same report, Amazon "disgruntled" publishers because of exclusivity demands, refusal to allow advertising, and revenue split that has been defined as "wolfish."[21] Actually, another report suggests that the situation may be even worse for publishers: a 35/65 split in Amazon's favor.[22]

Economic Issues and Management
Challenges for the Emerging E-book Industry

Once handheld ebook readers are developed that can successively compete with printed forms, two additional factors can be expected to come into play that could potentially encourage a transition away from print and toward the exclusive use of digital text. One is the ever-growing "go green" initiative that enjoys widespread backing from not only much of the public, but from business, industry, and government. A logical result would be an increasing economic incentive for making ebook systems work as viable substitutes for print.

The second factor is that once textual and graphic material is committed to a digital form, it can easily be used and reused in a variety of ways. The economic advantages of this are obvious. Once a text is digitized, it can be outputted in a variety of ways. Each time a new economically viable use is found for the same text, a new revenue stream is opened. In this manner, the potential for profit can, in some cases, be increased exponentially.

On the other hand, the Kindle's capability of converting text to audible speech for listening on the go opens up the potential for yet another ebook-related legal controversy: When a copyrighted textual ebook is converted to speech, is Amazon essentially reusing copyrighted material without permission? Should Google pay additional royalties in such cases?

When the types of material that once were only available through the paper-based print media become digitized by default, both consumers and entrepreneurs will see the economic and cultural benefits of the use of a digital form. We are poised to make written material available anywhere, with a level of ease and convenience never before imagined.

As far as periodicals are concerned, perhaps electronic delivery in an ebook form via a device like the Amazon Kindle could prove a more viable economic model than that followed by the typical newspaper website. Up to this point, newspapers have experimented, for the most part, with distribution of newspaper-like content through ordinary Web pages. Separate, discrete electronic issues of a newspaper could be distributed by wireless means to such devices as the Kindle.

This might prove more attractive to consumers and more lucrative to newspaper publishers than the Web page approach. If so, the value of ad revenue to newspaper publishers could be revived. Such issues could be offered on either a paid subscription basis or for free. Offering issues for free might still prove profitable, since high printing and distribution costs would be eliminated.

At the moment, as far as economic models for ebook and electronic periodical distribution and sales are concerned, the field is wide open for experimentation. Book publishers could offer the first chapter or two for free.[23]

Another possibility might be to offer a free ebook copy with the purchase of any print-based book.[24] Magazine publishers could offer eye-catching graphics and limited text as a means to entice customers to pay for complete access. Hints have surfaced that electronic delivery could be an effective way to distribute erotica. Ebook readers could allow such material to even be read by commuters on trains.[25]

One factor that should be obvious but, oddly, has been little discussed, is the potential economic impact electronic delivery could have on publishers' expenses. Electronic delivery of the former print media obviates dependence on traditional printing and distribution, with their attendant accelerating costs. Shipping costs would be eliminated. Expenses associated with storage and display in retail shops would similarly become a thing of the past.

An additional factor that has received but scant attention but which could prove vital to the success of the ebook concerns the elimination of concern over storage and weight. This could especially become a selling point for textbook publishers, since print-based textbooks are often bulky and heavy, and for professionals who need ready access to massive reference volumes.[26]

Use of ebooks, once they become widely available, could provide an answer to a matter of concern on many college and university campuses, that being the rising costs of textbooks. This concern has resulted in adoption of textbook rental programs at some institutions, while others are already experimenting with the use of ebooks.[27]

One distribution model, such as that employed by McGraw-Hill, allows the institution to purchase access to electronic textbooks for its students. Encryption of digital texts, as is employed by CourseSmart in its online textbooks, could eventually eliminate the used textbook market,[28] but might cause difficulties for distribution via some handheld devices like the Kindle.

On the other hand, the potential exists for professors to create open source textbooks, using readily accessible Creative Commons material. Depending on the course, other potential text material could be supplied from public domain Google Books.[29]

Use of digital forms coupled with highly advanced handheld reading devices could obviate any desire for printing-on-demand systems. Printing on demand constitutes, at present, only a tiny segment of book publishing. A POD publisher accepts a manuscript, advertises it on the Internet, then prints it only when someone orders a copy.

A variety of sources reported that sales of ebooks in 2008 accounted for only about 2% of total book purchases. According to one source, however, sales of ebooks rose 171% just during the period from September 2008 to a year later.[30]

The International Digital Publishing Forum and the Association of American Publishers (AAP) reported that U.S. retail sales of ebooks have jumped from around one and a half million dollars in the first quarter of 2002 to about $46.5 million in the third quarter of 2009.[31] Ebooks have been touted as, in the words of one observer, "the fastest growing area of the book trade."[32] Manufacturers of some models of ebook reader devices cannot keep up with production demands.[33]

Ebook sales have been on the increase at the same time that costs of traditional print-based publishing have been rapidly increasing. In some contexts, costs are rising faster than, according to at least one observer, traditional business models would allow.[34] Economic necessity may result in an accelerated push toward rapid acceptance of e-text distribution models.

For the moment, however, traditional print-based book publishers have been showing clear signs of anxiety over the apparent shift to digital. This is true for several reasons. A primary cause of their apprehension has been the unprecedented dominance of the handheld reading device field by the Amazon Kindle, coupled with Amazon's tendency to offer significant discounts for digital books as opposed to their print counterparts.

This has publishers worrying that they may end up being forced to lower prices on printed books as well. A senior editor of *Publishers Weekly* has been quoted as saying that traditional book publishers "view e-books and e-book readers as inevitable," yet they "don't want Amazon to be the only game in town."[35] In the meantime, for $9.99, Amazon offers popular titles that might ordinarily bring two and a half times that much in a printed edition.

One ploy that publishers have been discussing in order to ward off ebook-generated revenue drain is to delay the release of digital versions of print-media books by four months. A proposed alternative approach involves releasing an ebook version early, in an attempt to drive print sales.[36]

Still, very public rumblings of discontent on the part of publishers indicate a fundamental distrust of Amazon. Currently Amazon is offering ebooks at less than cost, and this is making publishers nervous.[37] The fear is that this sort of practice will result in an across-the-board reduction for all books, whether electronic or printed.

At the same time, textbook publishers are confronted with multiple options in terms of textbook delivery. Distribution to laptops and desktop computers offers the easiest encryption for copyright protection purposes. This could enable publishers to recoup what has been described as the "staggering amount of business that the publishers lose"[38] once used sales surmount sales of new textbooks. On the other hand, growing popularity of such dedicated ebook devices as the Kindle mitigates against distribution of ebooks designed solely for desktop or laptop use.

Another challenge for publishers is that their basic business model for funding new titles will likely need to be abandoned. Currently, print-based publishers pay authors in advance. They then ship copies to retailers on a sale-or-return basis.

This method, dependent as it is on continual short-term cash flow based on hard-copy inventory, is not amenable to adaptation to the ebook market.[39] In the world of print book sales, just because a book is shipped does not mean it is sold. The opposite is the case with ebooks. Ebooks might not generate much of a short-term cash flow, but many of the major expenses associated with print books are eliminated.

Generally speaking, publishers are not rushing, however, to embrace the new technology. For one thing, before publishers become enthusiastic about the new medium, some better provision for copy protection will likely be required.

When over a thousand book publishing industry professionals representing more than 30 countries were asked in 2008 if they were ready for the digital age, only about 70% answered in the affirmative. "Very few" of them expected ebooks or ebook reading devices to become the industry's primary revenue source before around 2013.[40]

Another thousand book trade professionals were queried in London in 2009. More than 67% of them said that industry professionals will require retraining in order to maximize use of digital media. Nearly 70% believed that essential to the growth of ebooks would be the ability to read them on various devices, although 42% said that most consumers will prefer to read ebooks on dedicated ebook readers.[41]

Similarly, consumers can only be expected to gradually adapt to ebooks. Currently, although Kindle owners tend to buy nearly double the number of ebooks as opposed to printed books, they continue to purchase the same number of print-based books as before.[42]

Conclusion

No one is certain what the future of the former print media will be like, but one thing is certain: The demand for written materials will continue. How those written materials will be distributed is not entirely clear. As current technological, economic, and social trends continue, however, we can expect the process of reading to increasingly shift from dependence on print to increased use of electronic forms.

The following are trends and expectations for the future that could potentially take what was formerly known as the print media into an era of exclusively electronic forms:

1. The gradual technical improvement of handheld reading devices and screen images;
2. The proliferation of handheld devices capable of direct input of commonly used file formats;
3. Growing recognition of the lucrative nature of digitizing text and graphics for multiple use;
4. Increased public acceptance of on-demand retrieval systems for text-based media;
5. Increased costs and decreased viability of such legacy media forms as printed books, printed periodicals, and microform;
6. Increased demand for environmentally friendly "green" alternatives to print-based media;
7. An industry shift away from dependence on one or two commercial entities (Amazon, Google Books) for ebook distribution generally;
8. Abandonment of ebook industry policies and practices (e.g., use of proprietary formats and necessity for fee-based file conversion) that do not encourage public acceptance of its products;
9. Increased recognition of the value of electronic features that enhance digitized textual documents, such as electronic document indexing, access to relevant Web links, and enhanced note taking ability; and,
10. Well-developed integration of ebook reader technology into handheld units capable of other media functions, such as video and audio playback, Web and e-mail access, and telephony.

As far as the last item is concerned, new tiny netbook computers sport screens that are small for computers, but larger than that of handheld ebook readers. These netbooks, of course, offer the distinct advantage of multiple capabilities and may catch on with book readers,[43] especially if screen quality greatly improves.

Electronic paper developer Joseph Jacobson of MIT has advocated that at some point in the future, each child will be presented with a digital book full of 240 pages or so of digital paper. He refers to this device as the "last book," because it will be capable of being filled and refilled with whatever text and graphics the user chooses. Because it can be reused indefinitely, traditional books will be rendered obsolete.[44]

The technologies involved in both publishing and reading have radically changed in a very short time and are bound to change far more. One factor can be expected to remain constant, however, and that is the public's demand for the type of media content that was formerly reserved exclusively for print. As one media executive[45] recently expressed it, the future "means believing that the word 'publish' is not bound between two covers, but instead defined by the art and craft of connecting great writers with their audiences."

Notes

1. Nicole Howard, *The Book: The Life Story of a Technology* (Westport, CT: Greenwood Press, 2005), 11.

2. Joseph F. Kelly, *The World of the Early Christians* (Collegeville, MN: The Order of St. Benedict, Inc., 1997), 104.

3. Elizabeth L. Eisenstein, *The Printing Revolution in Early Modern Europe* (Cambridge, England: Cambridge University Press,1983), 9.

4. James N. Green, "From Printer to Publisher: Mathew Carey and the Origins of Nineteenth-Century Book Publishing," in *Getting the Books Out: Papers of the Chicago Conference on the Book in 19th-Century America*, ed. Michael Hackenberg (Washington: The Center for the Book, Library of Congress, 1987), 26–44.

5. Frank Ahrens, "Gannett's Profit Plummets 53 Percent," *The Washington Post*, 20 October 2009; "Fat Newspaper Profits Are History," *Reflections of a Newsosaur*, 2008, newsosaur.blogspot.com, 21 October 2008.

6. Martin Zimmerman, "Story Not All Bleak for Newspaper Industry's Outlook," *Los Angeles Times*, 31 August 2009.

7. Matt Weingarten, "Writer Speaks to Industry's Decline," *Cosmos Magazine*, 2009, www.cosmosmagazine.com (19 February 2009).

8. Martin Zimmerman, "Story Not All Bleak."

9. Michael Kinsley, "Life After Newspapers," *The Washington Post*, 6 April 2009, A15; Sah Dong-Seok, "Saving Newspapers," *Korea Times*, 8 April 2009; "U.S. Bill Seeks to Rescue Faltering Newspapers," *Reuters*, 24 March 2009.

10. Jonathan Jones, "What You Should Know About the Newspaper Industry," *PBS Frontline Newswar*, 27 February 2007, www.pbs.org/wgbh/pages/frontline/news war (15 Dec 2009).

11. Jonathan Jones, "What You Should Know."

12. Eric Eldred, et al., Petitioners, v. John D. Ashcroft, Attorney General, 537 U.S. 186.

13. Paul Brown, "Ebooks Battle for Next Chapter: Publishers Are Now Willing to Embrace Ebooks—But Are They Ready to Head Off the Threat of a Format War?" *The Guardian*, 23 April 2009.

14. Tim O'Reilly, "Why Kindle Should Be an Open Book," *Forbes*, 2009, www.forbes.com, 23 February 2009.

15. O'Reilly, "Why Kindle."

16. Barrett Comiskey, J.D. Albert, Hidekazu Yoshizawa, and Joseph Jacobson, "An Electrophoretic Ink for All-Printed Reflective Electronic Displays, *Nature*, 394 (16 July 1998), 253–255.; Duncan Graham-Rowe, "Paper Comes Alive," *New Scientist* 179, no. 2414, 16–17; Johan Feenstra, Rob Hayes. *Electrowetting Displays*. Eindhoven, the Netherlands: Liquavista, 2006.

17. Stephan Füssel, *Gutenberg and the Impact of Printing* (Aldershot, Hampshire, England: Ashgate Publishing, 2005), 199.

18. Paul Brown, "Ebooks Battle."

19. Farhad Manjoo, "Fear the Kindle: Amazon's Amazing E-book Reader is Bad News for the Publishing Industry," *Slate*, 26 February 2009, www.slate.com (9 December 2009).

20. Jonny Evans, "Apple Tablet Will Revolutionize Ebook Publishing," *9 to 5 Mac: Apple Intelligence*, 30 September 2009, www.9to5mac.com (18 December 2009).

21. John Paczkowski, "Apple Pitching Tablet to Publishing Industry; Spring Launch Expected," *Digital Daily*, 9 Dec 2009, digitaldaily.allthingsd.com (9 December 2009).

22. Michael Pastore, "Will Amazon.com Conquer the Ebook Industry?" *Epublishers Weekly*, 5 May 2009, epublishersweekly.blogspot.com (12 December 2009).

23. "Read it and Weep: Why Ebooks Must Change the Record," *The Business*, 26 January 2008.

24. Rich Trenholm, "Top Publisher Predicts 'Ebooks Will Sink the Publishing Industry,'" *Cnet UK*, 7 April 2009, crave.cnet.co.uk (9 December 2009).

25. Katie Allen, "Heavy Tomes Turned into a Light Read with the Virtual Pocket Library: Publishers Expect Digital Ebooks to be the New Blockbusters," *The Guardian*, 10 September 2009.

26. Katie Allen, "Heavy Tomes."

27. Elizabeth Redden, "Toward an All E-Textbook Campus," *Inside Higher Ed*, 2009, www.insidehighered.com, 14 January 2009.

28. Wendy J. Woudstra, "The Future of Textbooks: Ebooks in the Classroom," *Publishing Central*, n.d. publishingcentral.com (7 December 2009).

29. James J. Duderstadt, "Will Google's Book Scan Project Transform Academia?" (online video), 2009, www.youtube.com (12 December 2009).

30. David Lieberman, "Tension Mounts in E-Reader Saga: Publishers Aren't Happy with Amazon's Pricing," *USA Today*, December 11, 2009, 1B-2B.

31. "Industry Statistics," *International Digital Publishing Forum*, n.d., http://www.openebook.org/ (9 December 2009).

32. Katie Allen, "Heavy Tomes."

33. Tony Bradley, "Kindle PDF Support Broadens Ebook Reader Appeal for Businesses," *BizFeed*, 24 November 2009, www.pcworld.com (5 December 2009).

34. Advanced Publishing, "Bosacks 5 Key Trends in Magazine Publishing—Clear Call to Action by Publishers," *Advanced Publishing*, 29 April 2008, blogs.advanced-publishing.com (26 September 2009).

35. David Lieberman, "Tension Mounts."

36. Kathy Gill, "Publishing Industry Responds to Digital Disruption by Delaying Ebooks," *MCDM: Flip the Media*, 2009, flipthemedia.com, 9 December 2009.

37. David Lieberman, "Tension Mounts."

38. Jeffrey R. Young, "How a Student-Friendly Kindle Could Change the Textbook Market," *The Chronicle of Higher Education*, 6 May 2009.

39. Rich Trenholm, "Top Publisher;" Evan Schnittman, "Why Ebooks Must Fail," *Black Plastic Glasses: Musings on Publishing and Life in the Digital Age*, 30 Mar 2009, www.blackplasticglasses.com (9 December 2009).

40. Marisa Peacock, "Ebook Publishing Trends: 70% Ready to Go Digital," *CMS Wire*, 15 October 2008, www.cmswire.com/ (9 December 2009).

41. Victoria Gallagher, "Bookseller Survey: Cheaper E-books Needed to Drive Digital Growth," *The Bookseller*, 12 February 2009, www.thebookseller.com (9 December 2009).

42. Farhad Manjoo, "Fear the Kindle."

43. David Rothman, "Trendspotting 2009: David Rothman's Predictions," *Publishing Trends*, January 2009, www.publishingtrends.com (26 September 2009).

44. Stephan Füssel, Gutenberg, p. 199; Joseph M. Jacobson, Barrett Comiskey, C. Turner, J. Albert, and P. Tsao, "The Last Book," *IBM Systems Journal* 36, no. 3 (1997), 457–463.

45. Lisa Holton, "Trendspotting 2009: Lisa Holton's Predictions," *Publishing Trends*, January 2009, www.publishingtrends.com (26 September 2009).

8

Challenges and Opportunities, New Models, and the Emergence of the Next Newsroom

Jen McClure

N EW NEWS MODELS ARE EMERGING AS traditional models break down. These new models are the result of developments in new media and communications technologies as well as a significant shift in the way that the public consumes and interacts with news and perceives its role in the journalistic process. Today's media and communications environment poses even greater challenges to newspapers than the introduction of broadcast media. For, while radio and television had a severe effect on newspaper circulation, the Internet has shown the potential to have a significant effect on traditional media models and to redefine our society's notions of journalism, news, community, and civic participation.

Budget cuts, editorial layoffs, declining advertising revenues and eroding circulation and readership are the challenges faced by most newspaper organizations in the U.S. Newspaper readership for Americans 34 to 64 years old is declining, in some cases at a faster rate than younger readers. Dubbing this the "Age of Indifference," the Pew Research Center for the People and the Press suggests that young people are not consuming as much news or in the same way as earlier generations. In 2002, Pew reported that even people who had become newspaper consumers had stopped.[1]

According to the Newspaper Association of America (NAA) and the Audit Bureau of Circulation (ABC),[2] the average circulation for U.S. newspapers is decreasing by approximately two to three percent per year. Additionally, 25 percent of newspaper jobs have vanished since 1990, and U.S. media employment has declined to a 15-year low. While newspaper jobs accounted for 50 percent of U.S. media jobs in 1990, today they represent only 38 percent.[3]

These statistics represent part of a steady decline that has been occurring over many years.

Meanwhile, the NAA also reports that traffic to newspaper websites continues to show impressive growth. According to a custom analysis provided by Nielsen Online for the NAA, newspaper websites attracted more than 74 million monthly unique visitors on average in the third quarter of 2009, more than one-third (38 percent) of all Internet users, Newspaper website visitors generated more than 3.5 billion page views during the quarter, spending 2.7 billion minutes browsing the sites over more than 596 million total sessions.[4] "Newspaper publishers continue to aggressively reinvent their business models, leveraging trusted brands to attract a growing and sophisticated audience in the digital space," says NAA president and CEO John F. Sturm.[5] And indeed it is true that the way we conceive and consume newspapers is changing dramatically. Even more impressive than the growth in visits to newspapers' websites is the growth of newspapers' blogs, which have enjoyed the largest increase in online newspaper readership by far over the past few years. Traffic to the blogs of the top ten newspaper sites increased more than 210 percent in December 2006 when compared to December 2005, garnering 3.8 million unique visitors in December 2006 alone.[6] But newspaper organizations are not currently meeting the public's primary needs and expectations. According to a July 2007 Harris poll, the top four priorities that citizens identified for newspapers were to better represent the communities they serve; to ensure that all points of view are represented; to provide more research and findings on key issues and raise the quality of writing and analysis. These priorities seem very basic, yet newspaper organizations are not meeting these needs and expectations. More than two-thirds of Americans (67 percent) currently believe that "traditional journalism is out of touch with what Americans want from their news," according to a 2008 study by We Media/Zogby Interactive.[7] The survey found that 70 percent believe that journalism is "important to the quality of life in their communities," but 64 percent are dissatisfied with the quality of journalism in their communities.[8] Respondents to this survey also believe that both traditional and citizen journalism is important for the future of journalism: with 87 percent stating that professional journalism has a "vital role to play in journalism's future," and 77 percent believing that citizen journalism is also important.[9] In addition, almost half of all respondents (48 percent) reported that the Internet is their "primary source of news and information," an 8 percent increase over just one year earlier.[10] Fifty-five percent of 18- to 29-year-olds rely on their Internet as their primary news source, compared to 35 percent aged 65 and older. Only 10 percent of all survey respondents stated newspapers as their primary news source, and just 7 percent of respondents 18 to 29 rely primarily on newspapers. An overwhelming majority—75 percent—stated that they believe the Internet has had a positive impact on the overall quality of journalism. And news delivered via the mobile Web is growing in importance. Nearly half (48 percent) of newspaper website visitors who own Web-enabled mobile devices are interested in receiving their news and information via these devices.[11]

These studies demonstrate the generational shift that is taking place. Digital natives rely more on online channels and alternative sources for their news than in the past, which presents news organizations with a complex challenge if they hope to continue to stay relevant and serve the needs of the public. News organizations need to focus on approaching journalism in a way that makes a difference in people's lives.

News Organizations and Google

"The *New York Times* is afraid of Google, and with good reason," notes Jay Rosen, a professor of journalism at New York University."[12] They can't lure the talent away from these Internet companies to reinvent the newspaper organization. Yet, Google doesn't want to be an editorial organization. If the company that figured out the Internet doesn't want to be an editorial company and the editorial companies haven't figured out the Internet, we have the makings of a crisis."[13] Yet, a large and increasing percentage of Americans rely on Google and Yahoo! to access news.

Patrick Phillips of Poynter interviewed David Eun, Google's vice president of content partnerships in 2008.[14] Eun, who formerly worked at Time Warner and NBC, now manages Google's relationships with traditional media organizations. While some observers see Google as a significant threat to the traditional journalism and news model, Eun asserts that Google is not a competitor to traditional media. "The biggest misconception is that Google has aspirations to become a media company, meaning that we would produce and own content that would compete against theirs. That's a major misconception. We don't produce or own content. In fact, we see ourselves as a platform for our partners that do."[15] Eun went on to comment, "I don't think they should be concerned about Google being a competitor. We're part of the Internet; we're part of the technological transformation that everyone's experiencing . . . We're not a media company, because we don't own or produce content. However, we work closely with those who do and work closely with advertisers."[16] When asked what is the biggest challenge facing most traditional media companies, Eun answered, "In traditional media you generally build your business by making your content scarce and bringing people to it. In digital media, though, the hurdles in distribution start to go away.

So your ability to get content out there is a lot easier. The bigger challenge for traditional media companies is to find an approach where ubiquity is the name of the game."[17]

But, despite Eun's characterization of Google as a valuable partner rather than a competitor to newspapers, newspapers' circulation figures and advertising revenues continue to decline. Many blame Google and question whether traditional news organizations truly understand how to make the cultural and technological changes necessary to operate effectively in the digital age.

The Emergence of Citizen Journalism

Another challenge to newspapers and the traditional journalism model is the emergence of "citizen journalism." The Internet has played an important role in the delivery of news since the 1990s. No longer limited to a daily newspaper and TV or radio broadcast news, many citizens now get their news from a far greater range of sources, including many websites that offer journalism in many forms. "Today, a new form of Internet journalism—citizen journalism—is taking root in which ordinary citizens are learning how to report on the people and events of the world with fresh eyes," notes Douglas McGill, a former *New York Times* reporter and Bloomberg News bureau chief.[18]

"YOU" declared the cover of *TIME* magazine in December 2006, and with that statement heralded a new era of media and journalism. In that issue, Rick Stengel, *TIME's* managing editor, wrote:

> There are lots of people in my line of work who believe that this phenomenon is dangerous because it undermines the traditional authority of media institutions like *TIME*. Some have called it "amateur hour" And it often is. But America was founded by amateurs. The framers were professional lawyers and military men and bankers, but they were amateur politicians, and that's the way they thought it should be. Thomas Paine was in effect the first blogger, and Ben Franklin was essentially loading his persona into the MySpace of the 18th century, Poor Richard's Almanack. The new media age of Web 2.0 is threatening only if you believe that an excess of democracy is the road to anarchy. I don't. . . .
>
> Journalists once had the exclusive province of taking people to places they'd never been. But now a mother in Baghdad with a videophone can let you see a roadside bombing, or a patron in a nightclub can show you a racist rant by a famous comedian. These blogs and videos bring events to the rest of us in ways that are often more immediate and authentic than traditional media. These new techniques, I believe, will only enhance what we do as journalists and challenge us to do it in even more innovative ways.[19]

In just the past few years, tens of millions of weblogs have been created, and now total more than 200 million worldwide, according to Technorati (www. technorati.com), a company that monitors what has come to be known as the "blogosphere" or the "live Web," tracking more than 112 million blogs. In addition, The China Internet Network Information Center states that there are more than 181 million Chinese bloggers.[20] In fact, although the growth rate of new blogs has begun to slow, the number has nearly doubled every five months for more than five years. The Pew Internet & American Life Project reports that a new blog is created every 5.8 seconds—or 15,000 new blogs every day.[21] In addition, since 2007, "microblogging" has emerged, due in large part to Twitter, which has, as of April 2010, more than 106 million registered users and is growing by 300,000 users a day.[22] These new, free, Web-based software products have enabled this "personal journalism" phenomenon, allowing citizens to have a voice, to engage in genuine and unlimited discourse and debate, to participate in the journalistic process by commenting on traditional news sources as well as creating their own form of journalism, alongside and parallel with mainstream media. Today the reader truly has become the writer, editor, and publisher—with the ability to begin to communicate and collaborate with other citizens to discuss, question, and begin to transcend the hierarchies and filter of traditional media channels.

Citizen journalism enables citizens to play an active role in the journalistic process by "collecting, reporting, analyzing, and disseminating news and information," according to what is considered to be a seminal report on the subject titled, "We Media: How Audiences are Shaping the Future of News and Information," authored by Shayne Bowman and Chris Willis and edited by JD Lasica.[23] This phenomenon is fueled by new communications tools and technologies, such as blogs, wikis, podcasts, online video, and Real Simple Syndication (RSS), which have laid the foundation for this emerging new form of journalism and participation by more citizens in the journalistic process.

The intent of citizen journalism, according to the report, is "to provide independent, reliable, accurate, wide-ranging and relevant information that a democracy requires."[24] Lasica classifies citizen journalism into several categories including citizen participation via commenting on news stories, creating personal blogs, photos, or citizen generated video, or local news written by residents of a community. He also includes independent news and information websites, participatory news sites and collaborative and contributory media sites, personal broadcasting sites and other kinds of "thin media" (such as e-mail and e-newsletters) in the definition of citizen journalism.

This new "personal journalism" emerged in the aftermath of 9/11, thanks to new publishing tools being easily accessible on the Web, enabling people to post immediate, dynamic, firsthand news reports with global reach. But more important, it happened because of citizens' desire to communicate, to share their stories, and to connect with a larger community during this crisis. The ability to "link," to literally connect stories, comments, updates, and varied perspectives to one another to create a web of conversation, community, and dynamic knowledge is critical to the success of this era of journalism.

In his essay, "Reweaving the Internet: Online News of September 11," Stuart Allen notes that many news analysts have called the attacks of 9/11 "the biggest story to break in the Internet Age."[25] Definitions of "news" and "journalist" were re-examined in the face of this crisis, driven by the need—and the technical ability—to communicate immediately and personally. Allen writes:

> Certain comments about "personal journalism" posted by readers of different web pages suggested that these forms of reporting may have provided some members of the online community with a greater sense of connection to the crisis than that afforded by the "official" news report.[26]

In stretching the boundaries of what counted as journalism, "amateur newsies" and their webloggers together threw into sharp relief the reportorial conventions of mainstream journalism.[27] Since 2001, we have seen this phenomenon of online citizen journalism continue to develop across the globe through many tragic moments, including the war in Iraq, the earthquake and tsunami in Indonesia in 2004, Hurricane Katrina in 2005, the student massacre at Virginia Tech, the California wildfires in 2007 and the Twitpic photo of the Miracle on the Hudson—the plane that landed safely in the Hudson River in early 2009.

However, until very recently many mainstream news sources with online properties did not judge these new news sources as valuable. As Allen notes, "By drawing on the vast array of information resources available across the Web, online news sites can provide their readers with background details or context to an extent unmatched by any other news medium. However, few of the major news sites in the U.S. made effective use of this capacity on September 11."[28] This lack of understanding about this important shift about how people wanted to access and interact with the news further fueled the decline of newspaper readership and the simultaneous growth of online citizen journalism to fill the need for information and commentary that was being unmet by traditional news organizations.

In 2002, Pew found evidence that a significant number of people who had been newspaper consumers had stopped reading them.[29] In fact, in some cases this has led to frustration on the part of some traditional journalists, who have made the decision to become citizen journalists. For example, Chris Allbritton, formerly of the Associated Press (AP) decided to leave his post with the

AP to solicit contributions on his blog, Back-to-Iraq.com, in order to provide him with the resources to conduct independent reporting on the Iraq war. He raised more than $14,500 from approximately 350 donors, which enabled him to go to Baghdad to post live, independent reports on his blog, free from any additional constraints that may have otherwise been imposed by his editors or news organization.[30] The donations covered his plane ticket, a GPS unit, a satellite phone, a digital camera, and money for food, shelter, and travel within Iraq. This is not only an important example of citizen journalism, but of a publicly supported citizen journalism initiative in action, one that provided Allbritton's readers and supporters with an alternative to the mainstream media's coverage of the war. His reporting drew more than 20,000 readers to his site www.Back-to-Iraq.com. Allbritton even allowed his sponsors to suggest his agenda. To his supporters he wrote, "If you'd like me to check out a story and it's physically possible, let me know and I'll do what I can."[31]

Meanwhile, trust in traditional media is at less than 50 percent in the United States, according to Edelman's Trust Barometer. However, that number is higher than in previous years, because "the definition of media has broadened to encompass social media,"[32] explains Richard Edelman, "and mainstream entities are now available as and when needed."[33] These societal trends and technological developments stimulated the emergence and the success of social media and citizen journalism, and also hold promise for traditional news outlets as they reinvent themselves by implementing new communications technologies within their own organizations and by encouraging more public participation in the journalistic process.

Jay Rosen, an active supporter of citizen journalism and what he has termed "networked journalism," commented on this development, noting that the whole dynamic of "professional journalists" is recent.[34] Journalism as a profession didn't exist until the twentieth century, meaning that there was not as much of a separation between the journalists and the public, or the professional journalism from the personal journalism and storytelling that was taking place earlier in history. "The dynamic of professionalization is devised precisely to separate the expert practitioner from the laypeople," says Rosen.[35] But the emergence of citizen journalism and social media have begun to change this dynamic. "None of us anticipated this. There's been a shift in power. When tools are distributed, that changes everything."[36]

Newsrooms Slow to Embrace New Media

Although the Internet allows for a high level of collaboration and immediacy, most news organizations until very recently still operated under a "print mind-set"—thinking of their communications to their audience as uni-directional, the journalist is the communicator and the public is the audience. Newspaper journalists and newspaper organizations have been somewhat limited by their medium—forced to treat news as a static documentation of what had occurred, as witnessed, reported and analyzed by the journalist.

The Internet, and more specifically the weblog, the vehicle used by many citizen journalists, is based on immediacy, links and multi-faceted communication. This ability to provide a personal perspective, an alternative viewpoint, to witness, collect, analyze, and refute traditional media's accounts, to link to original sources, as well as conduct up-to-the-minute, independent reporting are critical components of citizen journalism.

Citizen journalists use the Internet to share information and connect to others. They understand that the technology allows for a more participatory form of journalism. As stated earlier, the ability to "link," to literally connect stories, comments, updates and varied perspectives to one another to create a web of conversation, community and dynamic knowledge is critical to this type of journalism.

Meanwhile, until recently, many newspapers focused on simply using the Internet to record the news, rather than to foster communication about it. Initially, most merely transferred their old print model online, simply re-publishing articles from the print version of the newspaper, a practice that has come to be known as "shovelware." A new medium calls for a new model, but many newspaper organizations have not taken full advantage of its potential.

It is interesting to note that the dawn of the "Information Age" accompanied a decline in newspapers being relied upon as a primary source of information. In the early stages of the Internet, very few newspapers provided a true forum for dialogue, discussion and debate online.

Some journalists saw the opportunity and the threat of the Internet as early as 1990. In November 2007, a *BusinessWeek* article titled "A Cautionary Tale for Old Media"[37] recounted the story of former *San Jose Mercury News* executive editor Robert Ingle, who, in 1990 "envisioned that a global information network would emerge, giving rise to all manner of online communities."[38] Mercury Center was proposed to "extend the life and preserve the franchise of the newspaper." According to *BusinessWeek*, "Ingle laid out strategies for the entire chain: Give information to readers however they wanted it, integrate the print and online operations, and dream up new forms of advertising. 'I saw the Internet as a great opportunity, but also as a great threat,'" commented Ingle in the article.[39]

Over the past 18 years since Ingle's proposal, the *San Jose Mercury News* has felt the effects of that threat. Although it built a robust online presence, Ingle's vision of Mercury Center was never fully realized and the newspaper organization continued to function according to its traditional business model. As noted in the *BusinessWeek* article, if Knight-Ridder had embraced Ingle's model, perhaps the newspaper chain would have survived. But as advertising revenue continued to decline as the result of the shift to online advertising, Knight-Ridder was forced to sell the *Mercury News* in 2006. Since the sale, there have been multiple editorial layoffs, and the *Mercury News*' future is in jeopardy.[40]

According to *BusinessWeek*, "Executive editor Carole Leigh Hutton has vowed to 'blow up the newspaper' to make it relevant to today's plugged-in readers. . . . 'I'm figuring out the answers to daunting questions. What's journalism now? Who do we serve? And how do we make it work financially and journalistically?' says Hutton."[41]

But even after robust Internet tools became readily available later in the 1990s, few newspapers used them effectively to communicate with their audiences. A 2000 study of 120 newspapers' online entities found that less than 40 percent hosted online discussion forums. This represented no growth at all over a 1997 study. And many of those that did offer online discussion forums had very limited topics available for discussion and low participation by the papers' journalists.[42] The researchers concluded that this lack of participation by the journalists may have contributed to the low reader involvement and lack of public discourse. This suggests that journalists could have a significant role to play if they involved themselves more in conversations with the public, rather than just the transmission of news through their articles. The researchers ultimately concluded that while "the Internet has the potential to improve communication between journalists and audience, thus advocating democracy through user participation, media organizations do not necessarily exploit this opportunity effectively."[43] Another study conducted in 2004 compared online public journalism initiatives to those in the print editions of newspapers. The goal of the study was to examine the contents of online newspapers to assess if the use of the Internet supports public journalism in practice. The study used quantitative data to compare the content of online public journalism newspapers with online non-public-journalism newspapers, as well as the content of online public-journalism papers with their print versions. The researchers concluded that there was no indication that public journalism is better practiced online. "It is obvious that online papers can overcome many problems inherent in the offline media," according to the study.[44] "However, it is not easy to open up the potential of the online media. It is safe to say that most online newspapers have too limited staff and resources for the time being to be able to utilize the new medium's unique capabilities."[45]

Even in 2009, newspaper organizations were still only just learning how to harness the power of these new communications tools. According to Steve Outing writing on PoynterOnline, a publication of the Poynter Institute, a school for journalists, future journalists, and teachers of journalists, "A technique that too often is missing is the breaking-news blog. That is, when a major breaking local story hits, use the blog format to tell the story in bits and pieces as reporters learn the details."[46] He suggests that a "cultural shift" needs to take place in newsrooms. "The inclination to gather ALL the facts and then produce a story . . . is not the correct choice when it comes to stories that people need information about right away. That's old newspaper thinking that deserves to die. Some stories are important enough—are urgent enough—that they should be treated to a 'here's what we know now' and 'here's what else we just learned' an hour later treatment."[47]

Conversely, citizens seem to have easily understood and embraced the power of these new second-generation Internet technologies, often dubbed "Web 2.0," which include RSS-enabled blogging platforms, podcasts, online video and online communities and portals. The World Wide Web has been said to support a more "democratic" form of media because it provides citizens with access to news and information, the ability to share knowledge, question information, advocate, form online networks, debate ideas freely and even create their own media.[48] Web 2.0 tools enhance this ability.

New Tools and Technologies Lead to New News Models

Web 2.0, a term coined by author and Internet technology expert Tim O'Reilly, refers to technologies based on online communities and hosted services, such as social networking sites, blogs, wikis and folksonomies that facilitate collaboration and sharing between users.[49] These tools have not only enabled the citizen to become more involved in the journalistic process as a reporter and commentator, but also as an editor.

Why are these technological developments relevant to the public's relationship with the press? Web 2.0 has transformed how a large percentage of the public gets its news and information, and where it seeks conversation and community through media. For example, sites like Digg (www.digg.com), Fark (www.fark.com), and del.icio.us (http://del.icio.us/) allow users to aggregate, rank and rate news and information—empowering citizens as opposed to professional editors in a newsroom to determine what is important, interesting, timely and relevant.

Digg is a user-driven social content website. All content on Digg is submitted by its online user community. Content is then ranked by the site's users, and those stories that rank the highest are promoted to the front page. Every user can "dig" (promote), "bury" (lower rankings or remove), and comment on stories. Digg also allows its users to track other community members' activity throughout the site.

Similarly Fark is a news aggregator and an edited social networking news site. Every day Fark receives approximately 2,000 news submissions from its readership. In its first year (2005) Fark received 50,000 page views. The second year it reached one million. The highest-traffic corporate Internet hitting Fark's servers was CNN; the second highest was Fox News. Fark allows users to view breaking news from all major news sources on one site. It allows users to post, comment and vote on posts and links.[50]

del.icio.us (owned by Yahoo!) is a social bookmarking website. del.icio.us allows users to use "tags" or keywords to categorize online content. del.icio .us also allows users to see the links of other users and share links in return. Users can browse and search del.icio.us to discover the tags and bookmarks that others have saved. This results in a collaborative approach to labeling, categorizing and organizing news and other online information and content.

These are examples of some of the Web 2.0-based tools that are providing the public with alternatives to the traditional way of gathering, consuming, analyzing and sharing news, and, most important, allow citizens to actively participate in the editorial process by commenting on, ranking and categorizing the news. Today many newspaper organizations are beginning to incorporate these capabilities into their online offerings. In late 2006, the *New York Times* added links to social-bookmarking capabilities like Digg and Facebook to its online offerings. In addition, it has a branded channel on YouTube and an advertising partnership with Internet giant Google. According to Christine Topalian, manager of strategic planning and business development at NY-Times.com, the paper had been "looking for ways to tap a tech-savvy audience that is accustomed to commenting on and sharing news stories."[51]

In 2004, Jarah Euston, a 26-year-old Fresno, California native, founded Fresno Famous (www.fresnofamous.com), an online community portal with blogs, podcasts, Flickr photo galleries, and discussion forums. Euston wrote, "The community is very fragmented, and people might not know about local politics, music or events. Fresno Famous provides one place for everyone to feed on everything that we thought was good about the town."[52] The site's description of its mission and purpose follows:

> Ours is a community that is rapidly changing. How we change is up to us. . . . Want a city with cleaner air, better schools? Let's share our visions and talk about how we get there. Let's share news related to our goals. Let's comment on what our leaders are doing. Let's share our knowledge and engage each other in conversation.
>
> We know our users know more than we do. By pooling our knowledge, we'll discover and explore more. And who wouldn't love that? All the content on Fresno Famous is created by a combination of users and Famous staff. All users are welcome to post comments, stories, images, blog posts and more.[53]

Fresno Famous quickly became a hub for people to get informed about local cultural happenings, as well as city council meetings, school board decisions, and local politics. In short, people became more involved in their local community life via their participation in the Fresno Famous online community. In fact, after being in existence for only a short time, Fresno Famous played an active role in the discussion of downtown Fresno's redevelopment. In one instance, city government changed its plans for a development project as the result of Fresno Famous user complaints. "Fresno Famous influenced how the city thought about the issue," said Euston.[54] In December 2006, Euston sold Fresno Famous to the McClatchy newspaper conglomerate, which owns the *Fresno Bee*. Today the site continues to be a vibrant hub of community participation, and has even added more new social media features such as micro-blogging capabilities like Twitter, a free social networking and micro-blogging service that allows users to post very brief (140-character maximum) updates (or "tweets").

However, there does not seem to be any significant linkage between fresnofamous.com and fresnobee.com. In fact, fresnobee.com features its own blogs, community calendar, entertainment section and other features that Fresno Famous already does so well. This lack of integration of Fresno Famous content into the *Fresno Bee*'s online offerings seems to be a missed opportunity to engage Fresno Famous users and make the newspaper more relevant to this highly interested and participatory audience. This marriage between citizen journalism initiatives and traditional news organizations represents perhaps the most challenging but also perhaps the most promising example of the potential for a new model of journalism that combines citizen media with professional journalism in a collaborative online forum that adds to a multi-perspective, dynamically changing news environment.

The Emergence of the "Next Newsroom"

Our definition of "news" is changing as the result of new communications technologies. According to the "State of the New Media 2008" report from the Pew Project for Excellence in Journalism, news is becoming more "continual," thanks to new technologies like Twitter. In addition, the Pew report states that journalism must now "help citizens find what they are looking for, react to it, sort it, shape news coverage, and give them tools to make sense of and use the information for themselves."[55]

One such new news model is the "Next Newsroom" proposal. The Next Newsroom is an experiment—a proposal for a new model for journalism. But the proposal is not being developed solely from "within the craft" by traditional journalists and academics, but rather in consort with new media experts and citizen journalists as well. NextNewsroom.com, an initiative at Duke University, asks the question, "If you could build the ideal newsroom from scratch, what would it look like?" More than 200 members work collaboratively via the interactive site, a blog, a wiki and face-to-face events to pose questions and ideas and start experimental newsroom projects.

Another Next Newsroom initiative began in December 2006, when the Journalism That Matters Project (www.journalismthatmatters.org) issued a white paper and blueprint for the Next Newsroom.[56] The work, developed collaboratively by leading academics, journalists from traditional media companies, citizen journalists and emerging new media organizations, states, "This new ecology of news, powered by new digital tools, will be grounded in a partnership with citizens to build an informed local community."[57] The Next Newsroom blueprint sets out to "create a different kind of news gathering organization" based on four commitments:

1. To create an informed, engaged public as the best hope for the future of democracy and civic life,
2. To nurture healthy professional and citizen journalists who work together to produce reliable local news to serve healthy communities,
3. To create a local investment business model that sustains The Next Newsroom, without undue profit expectations, and with a goal of universal access to the news through digital delivery and other platforms,
4. To shape content through direct conversations and interactions with citizens, who retain oversight authority for the newsgathering operation. Control of content shifts from journalists to a mix of citizens and journalists who regularly meet, talk, share feedback. Management oversight of this news gathering enterprise rests with community directors, a community oversight board, or related community-based management structure.

The Next Newsroom paper also asserts that the foundation for the success of the Next Newsroom will be journalists who reflect their communities via a diversity of age, race, economic interests and politics; are that are "converts to citizen journalism" and who can bring different skill sets to the job, including valuing the contribution of the citizens and creating communities of interest and technological savvy—understanding the potential of the Internet and multimedia technologies.

The Next Newsroom white paper and blueprint proposes a different definition of news, a different relationship with citizens and a different approach to delivering content. The Next Newsroom bases its definition of news on conversations among citizens and journalists about what needs to be covered.

It suggests that Next Newsrooms host online conversations, use instant messaging, online forums and e-mail to gather feedback via kiosks and computer terminals in libraries, schools, retirement centers, etc. Additionally, editors would report to a community editorial board, thus shifting control of editorial decisions to citizens in the community. The Next Newsroom plan envisions the ability to provide universal access to news for everyone in the community, via laptop computers and mobile devices. This vision is, however, dependent upon ubiquitous broadband/Wi-Fi Internet access for every community.

The Next Newsroom is very much a work in progress. However, if the resulting Next Newsroom successfully incorporates new communications technologies and the underlying principles of public journalism, understands the value of including citizens in the journalistic process, and addresses the challenges of media literacy, education and access, the result will be a journalistic model that not only meets the American public's expectations but also is successful in sustaining itself.

The Next Newsroom plan developed by the Journalism That Matters group seems to suggest that the blueprint is designed primarily as a means for traditional newspaper organizations to transform themselves, and indeed we have seen some traditional news organizations use this to guide their recent transformations, as we will discuss later. But, as Jay Rosen has said, "We need news organizations, but they're not necessarily going to be newspapers."[58] And indeed, several new newsroom models have emerged in the past few years, including one initiated by Rosen himself.

NewAssignment.net is an example of a new type of journalism initiative. This online news site started by Rosen encourages citizen journalists and professional reporters to collaborate to "do journalism without the media,"[59] free from commercial pressures. NewAssignment.net's stories are developed openly and collaboratively on the site, via an open-source approach, rather than reporters and editors discussing articles in the confines of a newsroom. According to Rosen, NewAssignment.net's users are integrally involved in the creation of story ideas and assignments "that the regular news media doesn't do, can't do, wouldn't do, or already screwed up."[60] But NewAssignment.net does rely to some extent on professional journalists. "We think the hyrbid forms—mixing professional journalists and amateur contributors—are going to be the strongest forms, and we're attempting to show they have potential,"[61] says Rosen.

The project is not based on an advertising model, as with traditional news organizations, but rather was initially funded by foundations and corporate partners, and the articles are paid for through donations, which Rosen posits helps to encourage unbiased quality reporting that is relevant to readers. Rosen has coined the terms "networked journalism" and "open source reporting" to describe this new approach.[62]

Another example of a new news model is the City Journalist Directory (www.CityJournalistDirectory.com). Calling itself a "visual memoir in real time of worldwide current events, from major natural disasters to local happenings," the directory is somewhat like the Associated Press for citizen journalist material.[63] It allows citizen journalists to submit photos and videos of news in their local area from anywhere in the world. The City Journalist Directory makes citizen-generated news resources available by location and by event type (e.g., war news, natural disasters, local news, political events, etc.). City Journalist Directory states its mission as serving as a bridge between citizen journalists and the news media. Members submit photos and videos of news in their local area, and are paid if news outlets use their content. David Mitchell of the City Journalist Directory explains, "The City Journalist Directory was set up as not only a news resource, but also to serve as a sort of middleman between these individuals witnessing and capturing the news and the more traditional news media."[64]

If these new news models can be incorporated into newsrooms and newspaper organizations can embrace this more open and participatory form of journalism, they may be able to reinvent themselves and thrive. There are indications that this is beginning to happen in some traditional newsrooms. There is a symbiotic relationship beginning to emerge between some newspapers and the "live Web." Many newspaper organizations across the country are beginning to adopt new media channels in order to more fully engage with their audiences online. A 2007 study of America's top 100 newspaper websites by Bivings Research, titled "American Newspapers and the Internet; Threat or Opportunity?" indicates that newspaper organizations are increasingly using Web 2.0 technologies to expand their offerings and reach.[65] The research cites that 97 percent of newspapers surveyed now offer partial RSS text feeds, while three percent offer full text RSS feeds. Ninety-five percent of the papers in the study offer at least one reporter blog; 97 percent incorporated online video and 49 percent included podcasts. Ninety-three percent of the reporter blogs allow comments. This shows growth over 2006, when only 80 percent of the papers offered blogs, and 83 percent allowed comments. In addition, 24 percent included user-generated content; and 67 percent allow comments on articles. This represents a 14 percent increase over 2006, when only 19 percent of papers allowed comments on articles. Many large papers, including *USA Today, The Denver Post* and *The Washington Post*, also offer social networking on their site. This indicates a more conversational and dynamic approach to journalism than in the past and a growing understanding of the importance of online communities in today's society.

While some might suggest that many of these changes are no more significant than an updating of newspaper organizations' technologies and result in no more citizen participation than an online extension of the letter to the editor, the ability for reporters and readers to engage in discussion via a blog or social networking platform is in some ways fundamentally different than the way that newspapers have traditionally interacted with their audiences. Reporter blogs give the reporter a chance to explain the "backstory" of an issue and why they found it interesting or covered it the way they did. The ability for readers to comment provides them with an opportunity to ask questions, link the reporter's story to other sources, lend a different perspective or even refute the reporter's portrayal of the story or issue. These reader comments are not chosen or edited, as are letters to the editor. Nor are they static as are letters to the editor—these online discussion platforms are more immediate, more connected, have farther reach thanks to the global nature of the Web, have more permanence and can inspire more discussion and debate than a letter that appears once in a print publication that is thrown away the next day. This new medium allows for a "flow of social intelligence."[66]

In addition to utilizing new communications technologies more effectively, a cultural shift is beginning to take place inside newsrooms. Instead of viewing citizen journalists as adversaries or competitors or simply dismissing the value of their contribution, a significant number of journalists are discovering that social media can be good sources for leads, research and even breaking news that the mainstream media can pick up on and amplify. Other traditional journalists believe that bloggers complement their work in significant ways.

According to new research by the Society for New Communications Research, 70 percent of journalists are using social networking sites, 66 percent

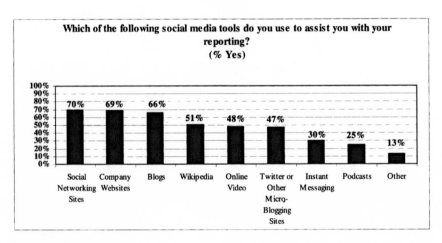

FIGURE 8.1
2009 Social Media Adoption Trends

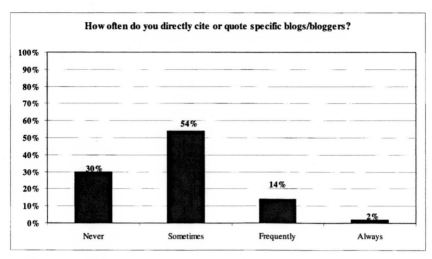

FIGURE 8.2
2009 Survey Findings

are using blogs, 48 percent online video, 47 percent Twitter and 25 percent are using podcasts to assist them with their reporting.[67]

Additionally, 70 percent quote bloggers in their articles, and 80 percent believe that bloggers have become important opinion shapers for the twenty-first century.

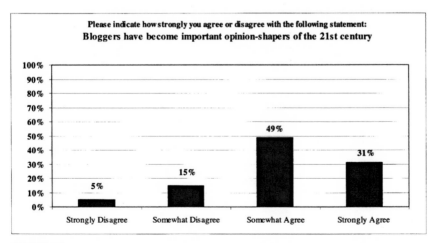

FIGURE 8.3
2009 Survey Findings: Perceived Value of Social Media

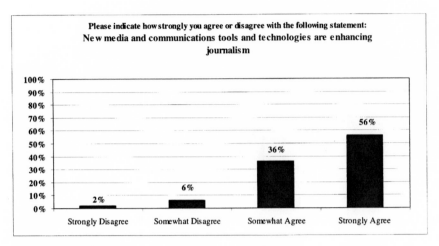

FIGURE 8.4.
2009 Survey Findings: Perceived Value of Social Media

Nearly all journalists surveyed (98 percent) agreed that new media and communications tools and technologies are enhancing journalism to some extent, and 87 percent feel that social media enhances journalists' relationship with their audience.

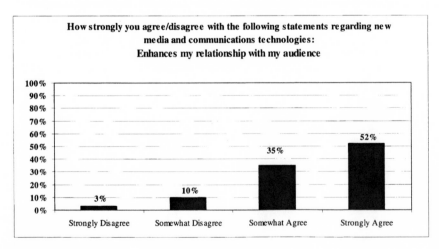

FIGURE 8.5.
2009 Survey Findings: Perceived Value of Social Media

What Lies Ahead for the Newsroom?

The changes taking place in journalism and the media landscape indicate the emergence of a new approach to newsgathering and a model in which news is not a static product, but rather in a constant state of evolution with a high level of citizen participation. The Next Newsroom is very much a work in progress. However, if what emerges as the Next Newsroom successfully incorporates new communications technologies, understands the value of including citizens in the journalistic process, and addresses the challenges of media literacy, education and access, the result could be a journalistic model that better meets the American public's expectations, and is therefore able to be more successful in sustaining itself.

While many newspaper organizations are transforming their organizations and approach to journalism, many others still do not understand that editorial resources are a crucial investment rather than just a cost or a line item on the budget. As a result they do not invest enough in the editorial resources necessary to adequately support their online or their offline initiatives or effectively communicate with their audiences. Most have not figured out the new economy of the Web, and some still do not value citizen participation in the journalistic process, nor do they believe that journalists should be actively involved in community building.

Additionally, even for those pursuing a more collaborative and dynamic news model, there are significant, overarching societal challenges, including ensuring equal access to educational and technological resources and ensuring that this new participatory form of journalism is not just a playground for the elite. Addressing these challenges calls for a new model and philosophy not only in newsrooms, but in American society as a whole, to ensure equitable access to news and information and that all voices are represented. News organizations and society as a whole have the opportunity and the obligation to ensure that our journalistic process continues to evolve and improve to better serve the needs of all citizens, but that will only be possible through access to educational and technological resources. Newspaper organizations must change their focus and their business models drastically—focusing not on print circulation numbers, or achieving unrealistic profit margins through editorial layoffs, but instead on initiatives to actively engage and empower citizens.

Newspapers must support literacy initiatives, especially media and computer literacy and address the digital divide, which is not just a third world issue but a problem that plagues this country today. They must support ubiquitous Wi-Fi initiatives in the communities they serve and champion network neutrality to ensure the free flow of news and information, unrestricted by telecommunications and ISP companies. Just as the post office was founded to deliver newspapers, today telecommunications companies are the conduits through which millions of citizens get their news. News organizations need to understand this and actively address this development. If, and only if, they can realign their priorities to address these societal needs, newspaper organizations will have an important opportunity to fill the news, information, and education needs of the American public and support the democratic process.

As *Fast Company* noted in its October 2006 issue, "Newspapers need to be the place where everything local is posted, shared, discussed, criticized, or mashed up. That means lots and lots of user content and very little 'publisher control.' This would be a significant philosophical shift for most newspaper management and editorial staffs."[68] Traditional news organizations have an opportunity to embrace the changes in the media and communications landscape to transform and enhance their role in American society. They have the opportunity to create a valuable forum for communication, public participation, debate, community building, and democracy—and to realize a successful business model as a result.

We have seen the newspaper change drastically and quickly over the past few years. However, in addition to disrupting the newspaper industry, this new era of online communications and increased citizen participation in the journalistic process also inspires us to address many questions. These developments are reshaping the way we view traditional journalism and the newspaper organization's role in society. In a larger sense, this new era of communications is reshaping and influencing the way we view, perceive, experience, and participate in our world. It invites us to rethink the difference between "news and information" versus "knowledge," to reinvent the role news organizations can and should play in society today and to examine our expectations of what it means to be a journalist and the roles we might play in the journalistic process and in our society as citizens.

Notes

1. Pew Research Center for the People and the Press, "Public's News Habits Little Changed By Sept. 11," June 9, 2002, 3.

2. Newspaper Association of America (NAA), http://www.naa.org (1 December 2009).

3. *Advertising Age*, March, 2008.

4. Nielsen Online and the Newspaper Association of America, http://www.naa .org/PressCenter/SearchPressReleases/2009/NEWSPAPER-WEB-SITES-ATTRACT-74-MILLION-VISITORS-IN-THIRD-QUARTER.aspx (1 December 2009).

5. Nielsen Online and the Newspaper Association of America, http://www.naa .org/PressCenter/SearchPressReleases/2009/NEWSPAPER-WEB-SITES-ATTRACT-74-MILLION-VISITORS-IN-THIRD-QUARTER.aspx (1 December 2009).

6. Nielsen/NetRatings, 2007.

7. "We Media/Zogby Interactive Poll," February 27, 2008, http://www.zogby. com/News/ReadNews.cfm?ID=1454 (1 December 2009).

8. "We Media/Zogby Interactive Poll."

9. "We Media/Zogby Interactive Poll."

10. "We Media/Zogby Interactive Poll."

11. "Power Users," Newspaper Association of America, 2006.

12. Quote from independent interview with Rosen conducted in July 2007.

13. Original interview conducted in July 2007.

14. Patrick Phillips, Media Interviews, In their Own Words, "David Eun: Google Won't Become a Media Company," I Want Media, January 30, 2008. http://www .iwantmedia.com/people/people70.html (1 December 2009).

15. Patrick Phillips, "David Eun," 2008.

16. Patrick Phillips, "David Eun," 2008.

17. Patrick Phillips, "David Eun," 2008.

18. Douglas McGill, "The McGill Report," http://www.mcgillreport.org/large-mouth.htm (1 December 2009).

19. Rick Stengel, "Person of the Year: You," *TIME*, 25 December 2006, 8.

20. Web2Asia, http://www.web2asia.com/2009/07/17/q2-2009-china-internet-statistics-report-released/ (1 December 2009).

21. Pew Internet & American Life Project, http://www.pewinternet.org/ (1 December 2009).

22. Dan Fletcher, "With 106 Million Users, Twitter Tries to Grow Up," April 14, 2010, http://www.time.com/time/business/article/0,8599,1982092,00.html (1 December 2009).

23. Shayne Bowman, Chris Willis, and J.D. Lasica, eds., "We Media: How Audiences Are Shaping the Future of News and Information," 2003, http://www.mabusite .com/tour/assets/we_media.pdf (1 December 2009).

24. Bowman, Willis, and Lasica, eds., "We Media," 2003.

25. Barbie Zelizer, Stuart Allan, and Victor Navasky, eds., *Journalism After September 11* (New York: Routledge, 2002), 127.

26. Zelizer, Allan, and Navasky, eds., *Journalism After September 11*, 127.

27. Zelizer, Allan, and Navasky, eds., *Journalism After September 11*, 130.

28. Zelizer, Allan, and Navasky, eds., *Journalism After September 11*, 130.

29. Pew Research Center for the People and the Press, "Public's News Habits Little Changed By Sept. 11," 9 June 2002, http://people-press.org/report/156/publics-news-habits-little-changed-by-september-113 (1 December 2009).

30. http: //www.back-to-iraq.com.

31. http: //www.back-to-iraq.com.

32. Social media refers to the various activities that integrate technology, social interaction, and the construction of words and images. This interaction is reliant on the varied perspectives and "building" of shared meaning, as people share their stories

and perspectives. Social media use the "wisdom of crowds" to connect information in a collaborative manner. Social media can take many forms, including Internet forums, message boards, weblogs, wikis, podcasts, photos and video, vlogs, wall-postings, e-mail, instant messaging, music-sharing, crowdsourcing, and voice over IP. Social media has a number of characteristics that make it fundamentally different from traditional media such as newspapers, television, books, and radio. Primarily, social media depends on interactions between people as the discussion and integration of ideas and shared meaning, using technology as a conduit. Social media is not finite: there is not a set number of pages or hours. The audience can participate in social media by adding comments or even editing the stories themselves. Content in social media can take the form of text, graphics, audio, or video. Several formats can be mixed. Social media is typically available via feeds, enabling users to subscribe via feed readers, and allowing other publishers to create mashups. Social media signifies a broad spectrum of topics and has several different connotations. http://en.wikipedia.org/wiki/Social_media.

33. Edelman Trust Barometer, http://www.edelman.com/trust/2008/.

34. Rosen, interview, 2007.

35. Rosen, interview, 2007.

36. Rosen, interview, 2007.

37. Steve Hamm, "A Cautionary Tale for Old Media," *BusinessWeek*, 6 November 2007, http://www.businessweek.com/magazine/content/07_45/b4057059.htm (1 December 2009).

38. Hamm, "A Cautionary Tale for Old Media," 2007.

39. Hamm, "A Cautionary Tale for Old Media," 2007.

40. Hamm, "A Cautionary Tale for Old Media," 2007.

41. Hamm, "A Cautionary Tale for Old Media," 2007.

42. Xigen Li, *Internet Newspapers: The Making of a Mainstream Media* (New York; Routledge, 2005), 242-256.

43. Li, *Internet Newspapers*, 243.

44. Li, *Internet Newspapers*, 243.

45. *Newspaper Research Journal*, Spring, 2004.

46. Steve Outing, "Grading Newspapers' Website Progress: B-," *Editor and Publisher*, 27 November 2006, http://www.editorandpublisher.com/eandp/columns/stopthepresses_display.jsp?vnu_content_id=1003438775 (5 January 2010).

47. Outing, "Grading Newspapers' Website Progress: B-," 2006.

48. Berkman Center for Internet & Society, Harvard Law School—http://cyber.law.harvard.edu/research/internetdemocracy# (1 December 2009).

49. John Musser and Tim O'Reilly, "Web 2.0 Principles and Best Practices," an O'Reilly Radar Report, November 2006, 4.

50. http://www.fark.com/farq/about.shtml#What_is_Fark.3F.

51. Cook, John, "New York Times Adds Sharing Tool: Readers Can Post to Digg, Newsvine," *Seattle Post-Intelligencer*, Monday, 11 December 2006, http://seattlepi.nwsource.com/business/295376_newsvine11.html (1 December 2009).

52. http://www.fresnofamous.com.

53. http://www.fresnofamous.com.

54. Jarah Euston, interview by Jennifer McClure, Palo Alto, CA, April 2006.

55. "State of the New Media 2008," The Pew Project for Excellence in Journalism, http://www.stateofthemedia.org/2008/ (1 December 2009).

56. http://www.mediagiraffe.org/wiki/index.php/Jtm-dc-next-newsroom.

57. http://www.mediagiraffe.org/wiki/index.php/Jtm-dc-next-newsroom.

58. Jay Rosen, interview by Jennifer McClure, Washington, DC, July 2007.

59. http://www.newassignment.net.

60. http://www.newassignment.net.

61. http://www.newassignment.net.

62. Jay Rosen, interview by Jennifer McClure, Washington, DC, July 2007.

63. http://www.cityjournalistdirectory.com/.

64. http://www.cityjournalistdirectory.com/.

65. "American Newspapers and the Internet; Threat or Opportunity?" Bivings Research 2007.

66. http://www.bivingsreport.com/2007/american-newspapers-and-the-internet-threat-or-opportunity/ (1 December 2009). John Dewey, *The Public and Its Problems* (Athens, Ohio: Swallow Press, 1954), 219.

67. "Middleberg/SNCR 2009 Survey of Media in the Wired World," research conducted by the Society for New Communications Research (San Jose, CA: SNCR Press, 2010).

68. *Fast Company*, October 2006, http://www.fastcompany.com/magazine/110/open_hyper-local-hero.html?page=0%2C3 (5 January 2010).

9

Broadcast and Cable on the Third Screen

Moving Television Content to Mobile Devices

Jennifer Meadows

TV on My Phone?

FOR YEARS WE HAVE BEEN WATCHING OUR favorite shows on our home television. Then we could watch TV shows on a computer screen. Now we can watch almost anywhere using a mobile device like a cell phone or digital media player. So, we can now watch television shows on the first screen—the television, the second screen—the computer, or the third screen—the mobile device.

Everyone seems to be carrying some kind of mobile device these days. You see people using them everywhere: in the car, on a plane, in class, at the movies, walking down the street, at the gym, at work, and even in bed! The mobile device has become a true personal media technology. Cell phones, including smartphones, digital media players (DMPs), and even portable gaming devices like the Sony PSP are increasingly being used to watch and enjoy television programming. This chapter will explore issues related to the management, distribution, and use of broadcast and cable television content on mobile devices.

Mobile Devices

The mobile phone is an extremely successful technology. According to the wireless association, the CTIA, 276.6 million people in the United States subscribe to wireless services, which is 89% of the population.[1] Mobile phones are no longer for just talking; people use them for all kinds of tasks such as texting, taking pictures, shooting video, listening to music, e-mailing and watching video.

A separate category of mobile phones is the smartphone. These phones, like Apple's iPhone and RIM's BlackBerry models, give users more services, including access to the Internet. Smartphones are more of a handheld computer than a phone. They are also characterized by their use of small apps—applications designed for a variety of uses such as checking the weather, playing games and reading the newspaper. Smartphone subscribership is growing at a faster rate than other types of phones. As of the second quarter of 2009, there are 26 million smartphone subscribers in the United States.[2] Leading smartphone makers include Apple, RIM, Palm, Samsung and HTC.[3]

One of the things that make new mobile phones and smartphones so useful is the cellular network. 3G networks allow mobile phone users to access data at broadband speeds allowing users to access the Internet as they would with a computer. For example, an iPhone user can watch videos on YouTube or purchase a pair of boots on Amazon.com (with the right apps) or a BlackBerry user can check scores on ESPN.com.

Digital Media Players (DMPs) are another extremely popular technology. Apple's iPod is the market leader with 73.8% of the market as of fall 2009. iPod classic and iPod nano users can purchase or rent videos at the iTunes store and iPod Touch users can also watch video via an Internet connection using the device's built in Wi-Fi and the appropriate app.

The Sony PSP is a portable gaming device that allows users to watch television shows through the PlaystationNetwork. The Sony network allows users to either purchase or rent videos and save shows on a Memory Stick. Video is also available through RSS feeds and with TiVoToGo, a service for TiVo DVR owners. Finally, PSP users can enjoy live television with LocationFree TV. With this service users have to have a LocationFree TV Base Station connected to their DVR or set-top box, the LocationFree TV software installed on the PSP, and a Wi-Fi connection.[4]

LocationFree TV is an example of a remote access technology. Another technology that allows users to get home television content onto a mobile device is the Slingbox along with SlingPlayer Mobile software. With this technology the user connects the Slingbox to a broadband network and the multichannel video programming distributor (MVPD) (i.e., cable) set-top box or DVR. Once the Slingbox is connected, all users need to do is install the SlingPlayer software (regular or mobile) onto a computer or mobile device. The user can then watch and control all the programming available to him or her at home. For example, you are in Detroit and want to watch the local news in San Francisco; with the SlingPlayer Mobile app or software on your mobile phone you access the set-top box at your home to watch your local game on the local channel. You can even access programming stored on your DVR.[5] TiVoToGo is a similar technology. TiVo users can save programming onto a computer and then transfer it to a mobile device.[6]

The Distribution Network

With mobile devices people can watch what they want to watch (almost), when they want to watch it and where they want to watch (most of the time). The challenge for broadcasters and MVPD providers is how to deliver programming to mobile device users.

First consider the distribution network or how the content gets to the device. In the case of mobile phones content can be delivered four different ways: the cellular network, MediaFLO, digital broadcast spectrum, and the Internet—through a wireless connection (most often Wi-Fi). As mentioned earlier, a 3G cellular network provides broadband speed for data services like Web surfing and video streaming. The four top cellular carriers in the United States, Verizon, AT&T, Sprint and T-Mobile, all offer 3G coverage that, in turn, allows high quality video streaming. Depending on the phone and service providers, any broadcast or cable content available to personal computer users on the Internet is also available via 3G cellular networks.

Another technology employed by wireless carriers is Qualcomm's Media-FLO, a technology for delivering multiple channels of television that is used in the United States by both Verizon and AT&T. MediaFLO uses a different part of the spectrum to allow wireless carriers to offer channels of television content without overloading their cellular networks with bandwidth-hogging video.[7]

Not to be outdone, television broadcasters can now transmit a mobile television signal using their digital television spectrum. The Advanced Television Systems Committee (ASTC) approved the standard, called ATSC-Mobile/Handheld on October 16, 2009. About 30 stations nationwide are broadcasting the signal as of Fall 2009 but the technology's deployment is expected to increase throughout 2010.[8] In order for this technology to succeed, though, mobile devices must to be able to receive the broadcasts.

Finally, Wi-Fi is another way cell phones and DMP users can access the Internet to stream or download television content. Some cell phones and DMPs allow users to connect to the Internet with a Wi-Fi connection. Then using a mobile browser or app, users can access streaming video online. In addition, some phones, DMPs and portable video game players can download television shows with a Wi-Fi or cellular 3G connection. These downloads can be free but most often are rented or purchased.

Figure 9.1 shows how different types of programming are distributed to mobile devices through a variety of channels. Programming can originate on both a national and local level. Local programming goes to the local broadcast television station and can reach the mobile user through the Web, an app or with a mobile DTV (ATSC-m/h) broadcast. Nationally originated programming can be for a national broadcast network, a local station through syndication, such as *Entertainment Tonight*, or for an MVPD. National programming for broadcast networks can then go to the local network affiliate and to the user through the local path or can bypass the local station and get directly to the user via a pay or rental download, a website, through MediaFLO, or an app. MVPD programming can reach a mobile user through a website, download or rental, MediaFLO or an app.

Users can obtain a wide variety of television programming. The selection depends upon the service provider, the network being used to access the programming, and the type of device.

Content: What's Available, How to Get it and How Can it be Monetized?

There is an astounding amount of video content available for mobile users, including everything from short YouTube video to full-length theatrically released films. This chapter will focus on the delivery of programming from broadcast television stations such as a local ABC affiliate, broadcast networks such as Fox and MVPD channels such as Discovery and ESPN. Using a smartphone or Internet-enabled mobile device any content available on the Internet may also be available to the mobile user. Local programming, as well as national network and MVPD programming ranging from local newscasts, soap operas, game shows, talk shows, prime-time comedies, reality shows, and concerts, to sports, is available. In other words, if it is on TV it is probably available to a mobile user. In addition, programming can be purchased or rented for download to mobile devices. There is less variety of programming available for purchase/rental and download but it will be available in a format that is designed for the mobile device.

Apps are also available for mobile television viewing on smartphones and some DMPs. An example of this is the CNN Mobile app that allows iPod Touch and iPhone users to watch live CNN news as well as on-demand programming.[9] Limited television programming is also available through cellular services such as Verizon's V-Cast.[10]

As discussed above, both live and on-demand content are available. Live streaming, apps and MediaFLO allow mobile device users to watch live television. The ability to offer live content is especially important for certain types of programming such as sports and news.

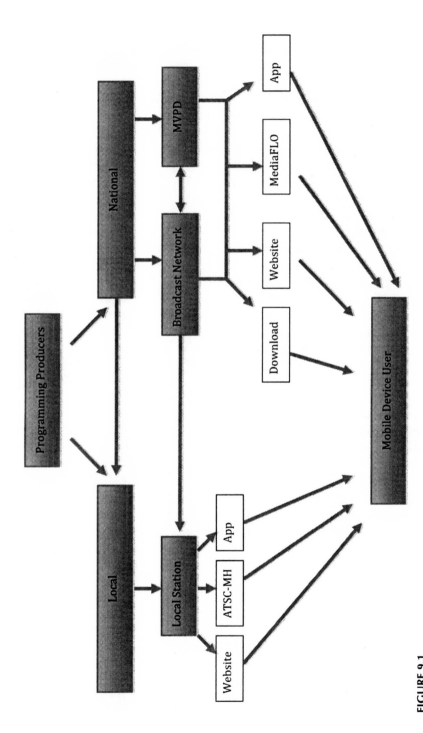

FIGURE 9.1
Equipment and Content Pathways

Websites

Available content depends upon the distribution method as well as the mobile device and the devices' operating system. One way broadcasters and MVPDs reach mobile users is through their websites. Local broadcasters, for example, stream local programming as well as network programming. Broadcast networks offer streamed full episodes online. MVPD service providers like Comcast offer full episodes on their websites.[11] MVPD programming providers (e.g., FX, Food Network) also offer full episodes online. For example, hulu.com, a combination of broadcast and MVPD interests, offers streams of broadcast and cable programming, exclusive Web-only content, and full-length movies.[12]

A major issue with viewing streaming video from a website on a mobile device is related to the popularity of Adobe Flash. The majority of video on the Web is played using Adobe's Flash technology. For example, to watch a video on hulu.com, mobile devices must have the latest version of Flash installed. However, the version of Flash that runs on computers is considered by many in the industry to be too slow and clunky for mobile phone operation. Adobe's Flash Lite, an earlier version of Flash for mobile devices was deemed inadequate by Apple, which still does not support Flash for its iPhone or iPods. Adobe announced in October 2009 that Flash 10.1 will work with mobile devices. So, Flash support is available or will be available by 2010 for other mobile operating systems including Windows Mobile, Android, Symbian and WebOS (Palm).[13]

Interestingly, Apple has not made any moves to incorporate Flash into their mobile OS. Why? There could be several answers. First, Apple has had a rocky relationship with Adobe, and Apple's own Quicktime competes with Flash. AT&T, the exclusive carrier for the iPhone, wants to preserve its cellular bandwidth and video streaming is a network-clogging activity. Finally, allowing iPhone and iPod users access to full television episodes online could and probably would cut into revenues generated by television episode sales and rentals on iTunes. That said, many iPhone users lament that they do not have a true Internet experience with the iPhone without Flash. Some apps have developed workarounds to allow Internet video viewing, but the ability to launch Safari and watch video as a user would on a personal computer is missing.

That "true" experience is currently available to some smartphone users and will be available to many more devices in the next year or two. For example, Adobe demonstrated Flash on the Palm Pre in October 2009. Users could watch video on numerous websites like Yahoo.[14] So, while Internet video streaming on mobile devices is not the primary way people now access television content on mobile devices it should only grow with faster processors in the phones and more advanced operating systems and software.

Downloads and Apps

Mobile device users can also download wide selections of television and cable content for purchase or rental. For example, iPod and iPhone users can purchase television programs at the iTunes store. These users can also purchase apps to watch mobile content. Most popular smartphones have apps for watching television but the iPhone has the most apps available of any smartphone as of fall 2009.[15] Apps are available for local and national television programming and for some Internet video services like Joost. In addition, apps are available to connect MVPD subscribers to their television content. For example, Comcast subscribers can get an app that allows them to watch selected video on demand (VOD) programming.[16]

MediaFLO

For a more traditional "channel-changing" television experience, cellular providers offer television services using MediaFLO. Verizon's V-Cast and Sprint TV are examples of these services. Sprint TV offers users channels of programming including the Disney Channel, the NFL Network, Bravo and the Weather Channel. Users can upgrade to the Premiere Pack and get NBC, CBS and ABC.[17] Verizon's V-Cast Mobile TV offers channels like ESPN, the Food Network, MTV and Comedy Central.[18] AT&T's Mobile TV service offers similar brands of television programming.[19] For these services subscribers need a compatible phone and pay an additional monthly fee.

Content Resolution and Audio Quality

An important consideration when discussing content is its resolution and audio quality. Several factors need to be considered including the resolution and audio quality of the digital media device and the content itself, not to mention the distribution method. Starting with the devices, some DMPs and smartphones were designed from the ground up with video in mind. For example, the iPod Touch has 480 by 320 pixel resolution, as does the BlackBerry Storm.[20] Other devices have lower resolutions. For example, the LG LX370 is 320 by 240 and the BlackBerry Pearl Flip is 240 by 320.[21] Audio is often inadequate on these devices because of the inferior headphones or earbuds provided with the device. Content is compressed for delivery and even though compression technology has improved over the years, compressed content still lacks the resolution and multichannel audio provided by broadcasters and MVPD services. Finally, depending on the bandwidth of the connection, content resolution and audio quality can change. Slower bandwidth can lead to stalled and glitchy streaming. To avoid this, users can always download the programming before viewing.

Revenue Models

Traditional

Broadcasters, programming distributors, and wireless companies generate revenue using several different models. The first is a traditional advertising-driven model taken from the traditional broadcast programming model. The programming can take the traditional route using linear mixed sponsor advertising inserted within the programming so the experience is similar to the television. Commercials can be placed at the beginning of the programming only. One company may also sponsor the episode, providing programming throughout the show. Online and mobile programming also allows control of commercial skipping and fast-forwarding. Users may not be allowed to skip commercials and may be asked to interact (for example by clicking or selecting) with the commercial to advance the programming.

Pay Per Click or Rental/Download

Another way to generate revenue is to charge per episode. The most well-known version of this is the iTunes store, where users rent or purchase shows. Most television shows are $2.99 for an HD version.[22] Season passes are available for purchase for many series—one fee and all episodes are automatically downloaded when they are available. Amazon Video On Demand also offers television programming in HD for about $2.99 per episode and users can purchase season passes. Amazon Video On Demand purchases can be moved to mobile devices using Amazon's Unbox video player. The device must support Unbox.[23] Sony PSP users can rent or purchase shows from the PlayStation-Network.

Subscription

Subscription-based services are another way to generate revenue. Some mobile device users pay a monthly fee to access programming. This is the case for V-Cast, Sprint TV and AT&T Mobile TV. Users pay a fee based on the programming package and get unlimited access to that programming. These services average $10 to $15 per month. There are even subscription-based apps like the *Wall Street Journal,* which used to be free. The paper announced in October 2009 that the app will come with a $2 per week charge for non-subscribers, and a $1 per week charge for subscribers.[24]

Paid Apps

Paid apps are a new way to generate revenue. Smartphone users can purchase apps for access to television content. For example, CNN Mobile costs $1.99 and is one of the top-selling apps on iTunes.[25] Apple reported in Summer 2009 that the iTunes app store was selling $1 million in apps every day. Apple keeps 30% of that revenue and the app developer gets the rest.[26]

Hybrid

Program providers can also use hybrid models, for example paid downloads with advertising or free apps with advertising. The opportunity to monetize content comes in many forms, none of which has yet been very successful. With Internet access on mobile devices and higher quality video streaming, users are increasingly turning to free television content over subscription-based and paid services. However, the advertising-supported model for the Internet has yet to generate profits. As of this writing Hulu announced that it is considering charging users a subscription fee as early as 2010.[27] Will people be willing to pay for something that has been free? In addition, widespread piracy in the form of illegal downloads and unauthorized streams are biting into potential revenue for content providers.

Advertising Challenges and Opportunities

Advertising on mobile media poses both challenges and opportunities. Google text ads have proven that measurable, personalized advertising works. With a mobile device, that device is attached to a person, not a household or a desktop. This presents an extraordinary opportunity for advertisers. To combine personalized advertising with mobile content creates a more robust opportunity for content monetization. In addition, many mobile devices like the Motorola Droid and the iPhone have built-in GPS. GPS provides an opportunity to connect mobile media with location-based advertising. For example, the mobile device knows the user watches the Food Network and if he or she walk by a Williams-Sonoma, store-related advertising and coupons can be automatically sent to that device, thus reaching the user. It is no wonder that Google purchased AdMob, a mobile advertising provider, in November

2009.[28] Mobile advertising does not mean success, though. As of fall 2009, it had yet to prove lucrative. Too much advertising might push users to pirated content.

Management Implications

Mobile media depends on relationships: relationships between device manufacturers and service providers, between content producers and service providers, and between users and service providers. Service providers like AT&T need to be sure that the mobile content they want to deliver can be delivered to their devices. AT&T mobile TV won't work without compatible phones. Therefore the wireless service provider needs to work with the handset maker to determine if the software and the network technologies will be compatible.

One framework for understanding these relationships is to examine the paths both the equipment and content take to get into the hands of the consumer. Table 9.1 shows these paths. First consider the consumer hardware: the mobile device. The device must be manufactured, distributed nationally and then made available to a local retailer. Now consider the content. The content production equipment must be made and distributed. The content has to be produced and distributed nationally and locally. In order for mobile media to be successfully adopted, the production and distribution functions of the content and equipment paths have to be fulfilled.[29] Mobile media content distribution is more complicated, because once content is created it can be distributed by a national distributor like a network, through a national mobile content distributor (service provider) such as Verizon or Apple, or even through a local distributor such as a local television station.

The relationship between the content producers and the service providers is critical. This relationship often depends upon protecting the content. In order to make money producing and delivering mobile content, that content often needs to be controlled. A key aspect of control is digital rights management (DRM) software. DRM can be used to protect content as well as a monetization tool. The technology allows rightsholders to limit access to and use of protected content. For example, DRM is necessary for subscription-based services like Sprint TV. In order to monetize the content through subscription services, DRM ensures that only subscribers get the television content. Most often DRM is associated with copyright protection. For example, a video purchased at iTunes cannot be shown on a BlackBerry or burned onto a DVD to play in a DVD player. It is important to note that the desire for protected content comes from the copyright holders and often not the digital content

TABLE 9.1
Equipment and Content Pathways

Equipment	Content
Mobile Device Manufacturing (Apple, RIM, Palm, Sony, Motorola)	Content Production Equipment (Sony, Grass Valley, Avid)
National Distributors	Distribution of Content Production Equipment
Local Retail Stores (AT&T Wireless, Best Buy, Walmart)	Content Production (ABC, Discovery, Fox)
	National Content Distribution (NBC, ESPN, HBO)
	National Mobile Content Distribution (AT&T, iTunes, Amazon, Sprint)
	Local Content Distribution (Local TV Stations)

seller. Apple and Amazon use DRM to appease the video content owners, thus allowing them to carry full selections of desirable television programs.

DRM is a contentious issue. DRM-free music is now available through major digital download sellers like Amazon and iTunes. DRM-free video is frustrating to users who want to enjoy purchased content on a variety of devices. However, content copyright holders want DRM to protect their content from piracy. There are efforts underway to consolidate DRM under one "roof." So, for example, a video purchased on iTunes could be viewed on other devices and a digital download from Disney could be watched on a PC, a DMP and a cell phone. Examples of this include the Keychest initiative from Disney and the Digital Entertainment Content Ecosystem, or DECE, that is supported by other movie studios, Sony, and IBM.[30] Digital content sales are fast becoming the norm for music but still lag in video. That will change.

For just about every DRM scheme, there is a DRM crack available on the Internet, though the effort and technical expertise it takes to employ these cracks, and break the law, is such that the majority of users put up with DRM. The relationship between the user and the service provider can help mitigate problems such as DRM. It is up to the service provider to give the user a positive experience with mobile media whether that be clear and error-free live television or seamless digital downloads. A Sprint TV customer needs to know when he or she goes to watch a show that it will be available and will run free of trouble. It should also be easy and intuitive to access those services. One way that Apple has been so successful with the iTunes store is that the process of shopping for, selecting, purchasing, downloading and viewing content is seamless. So while the freedom to purchase television content for an iPhone or iPod anywhere is limited, the experience that is available is incredibly simple. The user does not need different software and does not need to employ many steps; it just happens.

There needs to be a relationship between the content producer, the service provider and the handset maker. For example, consider apps for smartphones. A content provider makes an app like the E! online app. Apple has to approve the app and distribute it through its App Store and AT&T needs to provide the network over which iPhone users use this app. In another example, the content provided by Viacom for AT&T's Mobile TV has to be in a format that works with AT&T's handsets and runs over AT&T's MediaFLO network.

When television content moves across platforms, several factors must be considered. First the content provider must have a license to distribute the programming to mobile devices. The past writers union (WGA) strike over digital distribution royalties demonstrates the growing importance of this issue.[31] So for any company owning television programming it is vital to have the rights to digital and mobile distribution.

Second, entertainment companies and station groups can consider bundling content to create attractive programming packages for mobile users. For example, Sprint TV viewers could get a package of Viacom programming. While the programming all falls under one company, the variety of channels presented to consumers is large and attractive to subscribers: MTV, Comedy Central, Showtime.

Third, content providers can consider offering programming to mobile users as promotional items. For example, free HBO or FX show clips could entice consumers to watch the programming on their first screens (TV) and even sign up for premium channels. Thus, the mobile screen can be another venue for show and programming promotion and marketing. For example, Showtime has a free iPhone app that offers clips of Showtime programming and users can easily order Showtime and purchase other ancillary products.[32]

Conclusions and Future Implications

Table 9.2 provides a summary of the factors involved in bringing television content to mobile devices. As the table shows, the distribution network and technology, type of content, revenue stream and mitigating issues all factor into the success of the service. Mobile device use is growing and the success of cross-platform content means success for content providers like broadcasters and cable channels, device makers and device service providers.

First, issues surrounding mobile media delivery and content need to be addressed. The technologies that make streaming video work on the Web need to also work effortlessly on mobile devices. This means that versions

TABLE 9.2
Mobile Media Characteristics

Content Distribution	Internet Stream	Mobile App	MediaFlo	Mobile DTV	Pay per download or rental	Remote technology (Slingbox)
Live or on Demand	Both	Both	Both	Live	On demand	Both (stored on DVR)
Revenue	Ad supported	Free	Subscription	Ad supported	Paid	Device purchase
	Free	Paid				subscription
	Subscription	Subscription				
Mitigating Issues	Flash	Limited	Limited	Limited stations	DRM	Bypass service
	Network capacity	programming	programming	Limited	Limited	providers and
		Network capacity	Compatible	devices	programming	content
			device			distributors

of Flash need to work on and be supported by mobile device makers and/ or another means to view video cross platform needs to be developed and supported.

Second, the networks delivering mobile content need to provide enough bandwidth to provide a robust video experience to the end user. This means, at the very least, that 3G cellular networks need to be available to a larger percentage of the country. In addition, wireless service providers need to be looking at faster networks, such as 4G. As spectrum becomes available for new wireless services, these advances are becoming a reality.

Broadcasters should be considering adoption of the new mobile/handheld digital standard as this gives them another way to reach their audience and promote their programming. There needs to be a push to get device makers to incorporate the technology. How much promotion is going to come from a cellular provider like Sprint if the mobile broadcast competes with its own Sprint TV service? Partnerships between broadcasters and cellular providers can overcome this potential roadblock and, in turn, make mobile television more desirable and thus more monetizable.

One limitation of mobile devices is that they often do not work with other devices and/or services. For example, a television show purchased at Amazon will not work on an iPod. It is also often difficult to move content from one device to another and that frustrates users. Thus, DRM needs to be addressed. In addition, for wireless phones, different service providers may have exclusive deals to carry certain devices that only work with certain services. Black-Berry users cannot watch television from iTunes. Sprint mobile TV users do not have the same channels as AT&T users.

The Long Tail

The concept that media is becoming increasingly individualized and niche oriented discussed in *The Long Tail* can be easily connected to the personal mobile device. Users want to watch specific programming on their specific device when and where they want. Mobile media and mobile devices could be a perfect match. Chris Anderson's *The Long Tail* provides some good advice for "Long Tail Businesses," like mobile media.[33] Many of Anderson's suggestions for success can be applied to mobile media.

1. *Make everything available.* One limitation of mobile video right now is that programming is limited. While there is more and more available every day, the fact remains that mobile media users do not have the same selection of broadcast and cable programming available to them as first-screen and second-screen users.

2. *Help me find it.* Search technologies and interactive program guides are used effectively, but they could be expanded. Both Google and Bing have mobile search software. This technology needs to be better and more intuitive.

3. *One distribution method does not fit all.* Mobile television users can use multiple methods to access content, as shown in Table 9.2. The key is to make more content available across different distribution methods.

4. *One product does not fit all.* Mobile television is offered as a stream or as a download and even over the air. This variety should stay.

5. *One price does not fit all.* As discussed earlier, there are several revenue models for mobile television. Keep different price points for consumers who are willing to pay for advertising-free content and those who prefer to watch free advertising-supported content.

6. *Understand the power of free.* Free content can bring in users who will pay for better quality or exclusive content. Free previews of programming can entice users to subscribe to services. Free programming can also bring viewers from the third screen to the first screen. For example while waiting for a plane a person could watch a free episode of a show and after watching he or she may continue to watch at home, subscribe to the show or purchase other episodes.

People are watching more video than ever before and interest in watching television over mobile devices is growing. In order to capture this audience and evolve with the new media landscape, the media industry has to evolve as well. The audience wants what they want, when they want it, where they want it and how they want it. The media industry has to be able to offer that to the audience even if that means giving up old models of content delivery and revenue generation.

Notes

1. CTIA, "Wireless Quick Facts," http://www.ctia.org/media/industry_info/index.cfm/AID/10323 (22 October 2009).

2. Chris Quick, "With Smartphone Adoption on the Rise, Opportunity for Marketers is Calling," *nielsenwire* (2009), http://blog.nielsen.com/nielsenwire/online_mobile/with-smartphone-adoption-on-the-rise-opportunity-for-marketers-is-calling/ (22 October 2009).

3. Erick Schonfeld, "iPhone Makes Up 50 Percent of Smartphone Web Traffic in U.S., Android Already 5 Percent," *TechCrunch*, (2009), http://www.techcrunch.com/2009/03/24/iphone-now-50-percent-of-smartphone-web-traffic-in-the-us/ (22 October 2009).

4. Sony, "PlayStation.com-PlayStationPortable," (2009), http://www.us.playstation
.com/PSp (20 October 2009).

5. Sling Media, "Slingbox.com-Watch Your TV Anywhere," (2009), http://www
.slingbox.com/ (19 October 2009).

6. TiVo, "What is TiVo?" (2009), http://www.tivo.com/whatistivo/tivois/index
.html (23 October 2009).

7. Qualcomm, "Enabling the Convergence of Media and Mobile. The MediaFLO
System Overview Brochure," (2009), http://www.mediaflo.com/mediaflo/index.html
(15 October 2009).

8. Glen Dickson, "Mobile DTV Standard Approved," *Broadcasting and Cable*
(2009), http://www.broadcastingcable.com/article/358341-Mobile_DTV_Standard_
Approved.php (23 October 2009).

9. CNN, "CNN App for the iPhone," (2009), http://www.cnn.com/mobile/
iphone/ (23 October 2009).

10. Verizon, "Verizon Wireless: V-CAST Mobile TV," (2009), http://products.vzw
.com/index.aspx?id=fnd_mobileTV (16 October 2009).

11. Comcast, "Fancast," (2009), http://www.fancast.com/ (20 October 2009).

12. Hulu, "Hulu: About," (2009), http://www.hulu.com/about, (15 October
2009).

13. Ben Patterson, "Flash Support Coming to Android, Palm's WebOS, Black-
Berry; Where's the iPhone?" *Yahoo!tech* (2009), http://tech.yahoo.com/blogs/patter-
son/57474/flash-support-coming-to-android-palms-webos-blackberry-wheres-the-
iphone/ (22 October 2009).

14. Patterson, "Flash Support Coming to Android," (2009).

15. Yukari Iwatani Kane and Ryan Knutson, "iPhone Apps Take Root as
Cottage Industry," *The Wall Street Journal*, (2009), http://online.wsj.com/article/
SB125796886127143907.html?mod=rss_Today%27s_Most_Popular (11 November
2009).

16. Comcast, "Fancast," (2009), http://www.fancast.com/ (20 October 2009).

17. Sprint, "Sprint TV," (2009), http://www1.sprintpcs.com/explore/ueContent
.jsp?scTopic=multimedia192 (15 October 2009).

18. Verizon, "Verizon Wireless: V-Cast," (2009), http://products.vzw.com/index
.aspx?id=video (23 October 2009).

19. AT&T, "AT&T Mobile TV," (2009), http://www.wireless.att.com/learn/
messaging-internet/mobile-tv/ (16 October 2009).

20. Daniel Dumas, "Wired Smart Guide: Know Your Smart Phones," *Wired.com*,
(2009), http://www.wired.com/gadgetlab/2009/06/wireds-smart-guide-for-know-
your-smartphones/ (2 November 2009).

21. Mobiledia, "Cell Phone Comparison Results," (2009), http://www.mobiledia
.com/phones/compare/compare.php (10 November 2009).

22. Apple, "Apple iTunes: What's On," (2009), http://www.apple.com/itunes/
whats-on/#tv (5 November 2009).

23. Amazon, "Amazon Video On Demand," (2009), http://www.amazon.com/
Video-On-Demand/b?ie=UTF8&node=16261631 (30 October 2009).

24. Frédéric Filloux and Jean-Louis Gassée, "Why Media Firms Should Ignore 'Free,'" *CBS News*, (2009). http://www.cbsnews.com/stories/2009/11/09/opinion/main5587922.shtml (9 November 2009).

25. Apple, "Apple Reports Fourth Quarter Results," (2009), http://www.apple.com/pr/library/2009/10/19results.html (23 October 2009).

26. Yukari Iwatani Kane and Ryan Knutson, "iPhone Apps Take Root as Cottage Industry," *The Wall Street Journal*, (2009), http://online.wsj.com/article/SB125796886127143907.html?mod=rss_Today%27s_Most_Popular (11 November 2009).

27. Deborah Yao. "News Corp. Exec Sees Hulu Charging Fees For Access," *Associated Press*, (2009), http://www.google.com/hostednews/ap/article/ALeqM5iFFLwGfPgLLhFDm6nAiZNTzdp4RwD9BGE6S06 (5 November 2009).

28. Ian Schafer, "Why Google's Acquisition Of AdMob Isn't Just About Advertising," *Forbes*, (2009), http://www.forbes.com/2009/11/10/google-admob-schafer-cmo-network-schafer.html (10 November 2009).

29. August Grant, "The Structure of Communication Industries," in *Communication Technology Update and Fundamentals*, ed. August Grant and Jennifer Meadows (Boston, MA.: Focal Press, 2008), 52–65.

30. Brooks Barnes, "Studios' Quest For Life After DVDs," *The New York Times*, 2009, http://www.nytimes.com/2009/10/26/business/media/26stream.html?_r=1 (30 October 2009).

31. Dave McNary, "WGA, Producers Reach Tentative Deal," *Variety*, 2008, http://www.variety.com/index.asp?layout=festivals&jump=story&id=2476&articleid=VR1117980589&cs=1 (25 October 2009).

32. Viacom, "Showtime Mobile," 2009, http://www.sho.com/site/mobile.do (1 November 2009).

33. Chris Anderson, *The Long Tail* (New York: Hyperion, 2006).

10

How to Reach the Masses

Broadcasters' Use of the Internet and Cell Phones

Maria Williams-Hawkins

How to Reach the Masses

IN AUGUST OF 2005, Hurricane Katrina demonstrated what word-of-mouth communication could do. With limited electrical resources and pictures on television carrying the message of the devastation in New Orleans, people used impromptu meetings and whatever communication devices they could to raise money to help those left desolate by the hurricane. The oldest human form of communication, talking, generated millions of dollars in one week's time. The year 2010 opened with another devastating natural disaster, the 7.0 earthquake in Haiti. This time, social media was used to raise both awareness and money at a much faster rate.[1] CBS news reporter Daniel Sieberg reported that social media helped carry the message of the devastation on Facebook and blogs for those who had traditional computer access and satellite communication, while Twitter helped people share their feelings of sadness and anxiety about what they were experiencing within the first minutes and hours of the quake. Mobile giving helped those affected by the earthquake get dollars and doctors as well as outside rescue workers faster.[2] Twitpic helped the world see pictures of the demolished buildings and level of injuries sustained by thousands of people.[3] This time, celebrities were able to raise millions in hours.[4] Millions watched all these images around the world.

This chapter examines approaches taken by American television broadcasters to maximize their audiences by focusing on Internet-based programming options. The chapter provides a historical analysis of media approaches that led to current Internet-based technology and shows how these technologies provide programming options that compete with traditional broadcasting and also discusses the way traditional media uses Internet-based programming to enhance its viability. Internet-based media discussed in this chapter include blogs, Twitter and Facebook. The chapter closes with projections of how Internet-based media and traditional broadcasting may coexist in the future.

The Growth and Development of Broadcasting

Broadcasting began with the invention of the radio. However, inventing the radio was not a simple event. After international competitions to be the first with a successful model, competitions to develop the largest number of patents in the right combinations, radio was actually born. Whatever NBC, CBS and ABC could not agree on, RCA did agree that without programming, there would be no motivation to buy the radios that were prepared for consumers.[5] The practice of creating a product before creating a need has been the tradition in media. Inventors design and create new products. All they have to do is create a reason for consumers to feel that they "have to have" that product. New technology has been developed and the goal is to create programming that will fill the void and attract the consumer. Reaching the masses is an ongoing challenge.

Radio enjoyed a storied success during its era. Families stopped all their activities to hear their favorite radio show. Baby boomers may still recall listening to radio shows growing up. At that time, radios were priced reasonably enough for families to be able to afford one or more of a number of models within the family's price range. The development of the recording industry made programming radio stations less challenging due to the industry's novelty and especially when there was only one station in the area. By adding opportunities for local citizens to perform, talent was reasonably priced. The industry developed with the addition of advertising for local businesses and grew to its current stature. It was radio that brought Main Street into our homes.

With the success of radio, scientists were free to advance the technology to the next level. Inventors had developed a number of experimental televisions since the late 1800s.[6] By 1932, American homes could actually see a signal when W2XAB began broadcasting in New York City. The station was picked up in Raleigh, North Carolina, and reportedly seen in Bristol, Tennessee; Stevens Point, Wisconsin; Waterbury, Connecticut; Manhattan, Kansas; and, Adrian, Michigan, to name a few cities. [7] The Columbia Broadcasting System operated the station. Television was receiving fan mail. By 1945, when Westinghouse employees found out that broadcast signals could be broadcast from planes, technicians finally decided that "If it works, it will be revolutionary."[8] Television was ready to attract the masses. In 1940, experimental TV began broadcasting in Europe and the United States and scientists began developing color televisions in the early 1940s. So commercial broadcasting was ready to begin in 1941.[9]

No sooner than television was born, its demise was projected. Luddites said, "*This television thing will never last!*" At the time, it seemed that "they" were right. In 1942 until 1945 most television production was shut down due to World War II. Both television sales and television broadcasting were stopped. By 1946, the RCA 630-TS was on the market and the country's first color TV came out four years later, selling at the tremendous price of $499.95.[10] While radio had learned to reach the masses, television cost more and was struggling to have viewers who could afford to buy the expensive sets and programming that would keep viewers interested as well as sponsors who could afford to pay for the programming. In 1946, fewer than 7,000 television sets were in use.[11]

Just as television stole writers from radio and turned radio programs into television programs, today's new media has learned to do the same thing. When television was in its golden years, talent competitions worked well. Game shows worked well. Low-budget programs worked well. Television was able to take low-budget film productions and add them to their program line-up to augment the few productions that were made for television. An online *TV Broadcasting History* featured various articles from early viewers and industry representatives about what the networks broadcast in those early years. A report from Columbia Broadcasting System, which broadcast over WCBW, noted that they ran the test pattern from 7:30 p.m. until 8:00 p.m. and films from 8:00 p.m. until 10:00 p.m. A station director noted that they would run the same schedule as long as manpower and materials for broadcast prevailed.[12] According to him, at that time, having enough manpower was their greatest challenge.

Any production that allowed the audience to let the broadcasters know how much they enjoyed the programming was used. In those days, phone calls to the station and letters registered satisfaction or displeasure. Today's media reflects that same set of interests. More important, the lower-budget productions have often attracted the masses. Television programming since 2000 has included more game shows during prime time, more talent competitions that offer future careers for the winners and more reality shows that allow everyday people to talk about their personal challenges and get prizes for doing so. Families exchange mothers.[13] Overweight people try to lose weight together.[14] Nonprofessional dancers show how well they can dance[15] and people are taken to strange and exotic locations to live together for prizes.[16] More important, broadcasters at both network and local levels rely heavily on viewer response to these reality television shows. Broadcasters rely heavily on viewer comments from Nielsen, phone calls, letters and yes, Internet-based sources and social media.

During the early years of broadcasting, selling commercials or gaining sponsorship was imperative. The station with the greatest sponsorship made the most money. In addition to sponsorship, television stations relied on letters from the audience to let them know if they were offering the right programming for their audience base. As the industry developed, professional ratings services became available. Broadcasters got weekly reports of their ratings. Letters from viewers were still important but the statistical information that the ratings services offered gave broadcasters substantial information they could use to make their case for client support.

Reaching the masses was television's greatest skill for years. Radio was useful for times when people could not watch television or when they just needed noise. Television, however, did gain some competition from cable. Originally, cable television was only important in areas where the reach of traditional television was weak. Created in 1948 in the mountains of Pennsylvania by John and Margaret Walson, Community Antenna Television was not a big competitor for commercial television.[17] Cable TV served the needs of those who were in regions where commercial television was difficult or impossible to receive. Cable, however, slowly gained attention in the cities. Customers receiving cable service seemed to get better programming than commercial customers, or so they bragged to friends in the city. By the late 1970s, cable began to compete with television by showing targeted, syndicated programming and old movies. Cable networks like ESPN, Cartoon Network, Nickelodeon and Lifetime Television began in those years. Unlike network television, cable had taken the first steps to provide some aspect of a genre that was not routinely offered on network television. This was an early effort to go after a niche audience.

As the broadcast industry has grown, the quality of information that ratings services provide has improved in detail about the viewers as well as the speed at which that information can be delivered. Traditionally, demographic information such as age, gender and ethnicity was provided to ensure that advertisers could reach their population base and that programmers knew who to target with their latest hit production. Instead of getting monthly reports on their ratings, stations began to get their ratings within the hour and in some cases in 15-minute intervals. Broadcasters have to keep the audience around long enough to sell or expose them to products on the media platform the viewers are most likely using.

Formerly, broadcasters only competed against other broadcast or cable operations or family or personal activities. To increase its ability to compete, the size of televisions became smaller and smaller so that televisions could be taken along on any family activity and favorite programs would not have to be missed. Televisions moved from the floor-model fine-furniture items to seven-inch screens down to handheld models that could be watched discreetly when one was supposed to be engaged in other activities. Television was an important part of American life.

Based on the media hype today, one would think that no one is watching traditional television. Nonetheless, broadcasters continue to stay in the game. Many are programming the independent stations in their markets and carrying their news on more than one station. Most large and major market stations have separate digital channels that they program in a number of ways that bring the viewers back to the primary channel. Almost all are increasing the power of their news stories so that viewers have a reason to turn to their local stations instead of continually watching cable stations.

There is no less interest in broadcast or media-based messages, but there does seem to be less time for audience members to sit down and pay attention to their favorite show when it normally airs. Products such as On Demand[18] for Comcast Cable or HBO's On Demand[19] give satellite and cable viewers the opportunity to determine when they want to see their favorite programs. An alternative option is made available through In Demand[20] which offers the latest popular films as well as other programming to clients not affiliated with one specific cable or satellite provider.

Traditional broadcast programmers have returned to a 1950s programming approach and begun to rebroadcast popular programs more than once per week to attempt to give viewers a chance to see the newest episode and to cut the cost of programming. This approach is designed to attract new viewers as well as keep the ones stations had during less competitive times.

Unwilling to give up, network broadcasters have even decided to play with the competition. With changes in ownership rules, the same conglomerate that owns some of the cable stations may also own network affiliate stations. Therefore it is not uncommon to hear talent on network programs direct viewers to cable stations that are in the same family of stations. Local programmers take into consideration that their network programs may also be on the cable and independent stations in their area. Since the networks have agreed to sell off their programming, the local stations have no choice but to follow suit. Most make sure their local newscasts are on the independent stations and recognizing that "resistance is futile," their news is online as well.

Programmers are facing the biggest challenges of many of their careers. In interviews with large market programmers, several said that their jobs have changed from mildly competitive to extremely stressful as they attempt to put the right shows in place against the local as well as network competition.[21] Once they began to factor in online or Internet-based programming their jobs really began to change.

The goal for broadcasting in general had always been to reach the largest audience possible. What had been an effort to reach a massive audience changed to getting the largest audiences on multiple platforms. Programmers wanted the largest number of viewers to watch their news when it was broadcast on their station as well as when it was broadcast on the independent station(s) as well as when it could be watched online and where possible, when their news reporters could be seen in airports, sporting events or other venues.

Who's Going to Pay for This?

For years, broadcast station managers would tell you to never trust any broadcaster who told you that they just wanted to produce good programming. Most would say they entered television to make money. Recognizing that someone had to pay for "good programming," they knew that salesmanship of commercials and sponsored programs kept stations on the air. Radio brought the hardware store and the dress shops into our homes. Television showed us the specific products they were selling and how to wear or use them. Broadcasters set up our media-based sales expectations. Next to word of mouth, broadcasting was our best bet.

Now, however, stations sell on multiple platforms. Based on their ownership structure they may sell television spots primarily but might offer packages that include radio station spots as well as some Internet-based spots. The most challenging part of this endeavor is that the commercial breaks that air on the broadcast channels do not work for the online programming. Loss of commercials means loss of profit per show. Commercials and sponsorship are imperative for the broadcast industry. In an effort to be profitable, the length of time afforded for commercial breaks changed from the standard two minutes per break to three minutes or more. Additionally, commercial length changed from the standard 30-second spot to 10-second spots that might run three times in one commercial break period. For online programming, stations sell banners, 10-second spots and, in a few cases, 30-second spots before viewers get to see the actual programs.

Since 2007-2008 with the writers' strike,[22] broadcast programming has been in a state of flux. Trying to keep programming on network affiliates when new scripts were not being produced wreaked havoc on the industry.[23]

When there is not enough new programming, attempting to sell commercials becomes problematic. An already depressed industry saw itself running more reruns and more infomercials when regular programming could not be aired to generate better profit. Some stations began to run more one- and two-minute extended commercials while others ran more half hour infomercials. The use of the extended commercials gave viewers more and more two-minute opportunities to be convinced they should get police, technical support or medical training in their spare time. They were taught to get rich in the real estate business or build bodies like Adonis in 30 minutes per day. These low-budget commercials fill in the profitability for the day but do not provide the level of support stations need. They also fail to attract viewers who want to see the whole show or to watch the show more than once. While infomercials are not the greatest audience generators, they do attract more viewers than dead air. Infomercials tend to air during times of low viewership, overnight and on weekends. Viewers have become accustomed to them and know what to look for They also give viewers a real reason to seek programming elsewhere, often on   different platform.

 ̄raditional broadcasters are having a tough task of keeping their stations goi..g. During prime time, shows with high ratings are being cancelled because it costs too much to keep the shows on the air. Some characters are being written out of scripts because the actors' salaries are so high that the shows' profitability goes down. Other shows are returning the same characters but without raises.[24] Lots of new shows are being created for shorter periods of time to see if they can attract audiences before production companies invest in the cost of producing thirteen episodes. Cable networks are airing some shows for multiple times each week and creating only eight to twelve original episodes. Cable stations have chosen to create more original programming even though they produce fewer episodes.

Finally, prime-time as well as daytime programmers, are submitting to the need to tie themselves to the technological changes that are not going away. The *Today Show* has been airing competitions that ask viewers to vote to select the winners through their cell phones. *Today with Hoda and Kathie Lee* has regular segments built around the audience e-mails that arrive while the show is on the air. *The Ellen DeGeneres Show* has a regular segment that allows the audience to respond to questions or phone in their opinions about the topics for the day. Even Dr. Phil tweets certain viewers with personal questions. Prime-time programming often mentions e-mail, Facebook, Twitter and other Internet-based products. So, the cat is out of the bag—broadcast and Internet-based media can and do coexist. In the next section we will discuss the various ways the two technologies manage to coexist.

Where Did This Show Come From and Who's Watching It?

Although television's demise has been predicted for the past sixty years, during the last five years, the belief that "the end is near" has seemed stronger than ever. Article after article decried the end of television. Prognosticators argued about how and where we would get our news. Some projected that we would all turn off the evening news because it does not come on when we want to see it. Time seemed to be the key factor. The sag in the economy also contributed to a very robust belief that almost all was lost. Television stations began combining news departments around the country.[25] News departments in Ohio, Indiana, California, Hawaii and Illinois used various approaches to collapsing or consolidating news departments. In the case of the Chicago-based Tribune Company, four of their television stations in Miami and Hartford were combining their news crews in order to reduce operational costs. There would be one set of reporters to cover every event instead of having representatives from each station present.[26] Networks began cancelling shows on their premiere nights if their ratings were not high enough. Moreover, networks began cancelling television programs with good ratings when the overhead became too high for the companies to gain the profit they wanted. The constant barrage of reports of doom and gloom left industry practitioners baffled and arguing.

Yet, a November 10, 2009, story in *MediaPost* News alerted readers that Americans were watching more television in 2008-2009.[27] According to Friedman, Americans watched 4 hours and 49 minutes of TV per day in 2008-2009. Even though the development of DVRs had been identified as the reason people would not watch television, Nielsen's report indicated that DVR use added seven days of recorded viewing. While these figures only represent a 1.4% increase in viewership, a four-minute increase, viewership was up in dayparts other than prime time. Additionally, an October 26, 2009, *Media Post* News article indicated that children's viewing of TV was at an 8-year high. "Children between the ages of 2 and 5 spend 32 hours in front of their TVs."[28] Pre-teens, however, do not consume as much television as the younger children. Research shows that they are using other media to supplement their traditional television viewing. Friedman concluded "this trend mirrors overall media consumption in the two years. TV viewing is higher, and so are Internet, gaming and mobile phone activities."[29] From an advertising perspective, Nielsen's research shows that younger children are watching more commercials than are older kids. Younger children watch more commercials than any other demographic group. Based on the May 2009 sweeps, 2-5 year olds watched 50% more commercials in DVR playback during prime time based on research on the four networks.[30] Other statistics from Nielsen suggest that viewership figures are similar for teens and adults. So, perhaps the demise-of-television prophecy is still premature.

Statistics in 2009 on television use and TVs in the household should make broadcasters feel a bit more comfortable. Nielsen reports that in November 2009, almost 115 million American households had at least one television. Almost 104 million homes were either cable or satellite ready. Of those homes that had cable, 51% had digital cable. About 33 million homes have satellite TV and 47 million homes are HD capable. Indicating some progress, the number of homes using VCRs has gone down by 10% since 2007 while DVD player use is up 1% to 101 million. DVR recorders are up 12% since 2007 to nearly 37 million in use.

As for television viewing, broadcasters can breathe a sigh of relief when reviewing Nielsen's end-of-2009 statistics.[31] More than 99% of video content watched is watched on traditional television sets. The average American watches nearly 32 hours of TV each week. DVR users are watching almost eight hours of recorded TV each month. That's almost a 23% increase since 2008. And reflective of other data, children between the ages of six and eleven watch 28 hours of TV each week.

Why Do Viewers Watch?

Trying to reach the masses certainly would require broadcasters and new media producers to know the answer to that question. For traditional researchers, an examination of the Uses and Gratifications theory[32] would explain why viewers watch what they watch, while Everett Rogers' Diffusion of an Innovation[33] might better address the use of the newer media options.

Looking at the 1940 uses and gratifications theory, we recognize that the media have always competed against other sources of gratification (i.e., doing other things that please us as well as us watching, listening or participating in other media). Additionally, gratification can be derived from watching a favorite media show, watching a program in one's favorite genre and/or watching a show with those one loves.[34]

The media can provide information, programming that fits one's personal identity, integration and social interaction and entertainment. Both traditional and Internet-based media provide information. Information is provided through news-based programming, advice sources, self-education options and information provided in programs that build one's feelings of security. Media help build personal identity by presenting programming that reinforces personal values, showing models of various behaviors and providing programming that offers insight into one's self as well as showing characters with whom viewers can identify. Media provide integration and social interaction when programming presents characters that generate social empathy. Use of the media offers opportunities for feeling that one belongs to the group, provides options for conversation, creates a substitute for real-life relationships and allows users to connect with family, friends and society. The media offers entertainment frequently through its escapist, relaxing, cultural, emotional and sexually arousing programming. Based on these factors, it is easy to see that both traditional and Internet-based media meet these needs. In the next section we will discuss the problems broadcasters face in this changing mediascape.

The Challenges of Traditional Media

For years, television met the needs of most people. If viewers needed to find answers to questions or find validity in their behavior, characters and experts on television offered examples of ways to handle problems. The creation of new media caused challenges for television. Competition created a feeling of loss of the audience base. A September 24, 2009, article in *Media News International*,[35] reported a story that Americans are using more online and radio sources for news and information than TV or newspapers. The study, sponsored by ARAnet and conducted by Opinion Research Corporation, indicated that television use dropped by 3.6% while radio increased 2.9% and online usage increased 1.9%. Only 31% of those surveyed said that they get their news and information from television.

In addition to less than exciting numbers, the Opinion Research Corporation (ORC) survey indicated some changes in demographic use that is threatening to traditional broadcasting. The big four networks have worked to improve their ethnic programming and viewing base. ORC's research suggests that ethnic and medium income groups are more frequently using Internet-based media for news and information programming. Specifically, Hispanics, college-educated people and people earning more than $100,000 per year were increasingly using more Internet-based sources than the general population.[36] These findings were supported by one of *Advertising Age*'s Power 150 Bloggers, Dave Fleet.[37] Fleet indicated that his survey showed that high earners and college graduates are turning to Internet-based media more frequently. Supporting other research, however, Fleet did agree that traditional media was not "going away any time soon."

Internet-Based Media

Social media is a commonly used expression that has numerous definitions. For the purposes of this chapter, social media and online media fall into the category of Internet-based media. The terms will be used interchangeably. Discussions with technological experts suggest that social media has been around longer than most people realize. In an e-mail communication on October 12, 2009, with Brad King, a social media observer, he stated, "Social media has been around for about 50 years. It's only a big deal because everyone is using it now." Online technologies are those that allow people to share opinions, insights, experiences, and perspectives with each other.[38] Other sites identify social media as being user-generated, using social networking sites such as Digg, Reddit and Facebook, among others.[39] Anvilmediainc describes social media as an umbrella term that integrates technology with social interaction and involves the construction of words and pictures.[40] Sprythink.com brings in the addition of mobile technology as part of the technology used to facilitate discussions between and among people.[41] Social media are works of user-created video, audio, text or multimedia that are published and shared in a social environment, such as a blog, wiki or video hosting site.[42] These media use software tools that allow groups to generate content and engage in peer-to-peer conversations and exchange of content (e.g., YouTube, Flickr, Facebook, MySpace, etc).[43] Social media is the collection of tools and online spaces available to help individuals and businesses to accelerate their information and communication needs.[44]

Measuring Social Media

Just as the use of the Internet has expanded, so has the use of social media. Statistics compiled in 2009 identified the many ways that social media has affected American culture as well as the media. That year the U.S. Department of Education announced that students who were taught online outperformed students taught face-to-face. In higher education, 17% of all students are enrolled in online classes. That year, the second largest search engine was YouTube. Social media use has overtaken pornography as the number one activity on the Web.[45] A *Wall Street Journal* article suggests one in eight couples married in 2008 met online. In evaluating the statistics related to blogging, mobile media and other Internet-based operations it becomes apparent that social media has affected the population.

While broadcasters try to determine which social media to affiliate, they also need a scheme for determining which media is most appropriate for their message and product. MediaPost columnist David Berkowitz developed 100 ways to assess the value of the social media you use.[46]Although his approach was not scientific, it did include some valuable approaches to assess how well some people are using or not using certain media. Although not stated explicitly in each listing, it's clear that any effective social media used will generate a lot of discussion. The word "buzz" or talk is used in nine out of the first ten suggestions for measuring the effect of social media.

Some entries here can be interpreted several ways. Depending on how you define them, some of these metrics may seem redundant, while others may seem so broad that they can be broken out further. Many of these can be combined with each other to create new metrics that can then be tracked over time. The categories are: buzz, fan base, technological reach and popularity.

In analyzing the use of Internet-based media and traditional media, readers will find a strong correlation between the categories that Berkowitz created and the way that traditional broadcasters use social media. In the following section, we examine the historical development of social media including blogging, Twitter and Facebook. The section discusses the various approaches to blogging that fulfill programming expectations.

Before There Was Blogging

Discussion of blogging harkens back to research on communication within communities dating back to Webber in 1963.[47] Research on computer-mediated communities brings us closer to what we see in Internet discussions and mobile rants. Early discussions may have focused on citizen democracy, giving voice to those who felt their opinions were left out of the national conversations. While current analysis of these virtual communities shows that such conversations still have their place, there is now a movement past the political to business and topics as mundane as how to handle everyday problems.

In this section we explore the various historical paths identified as the "true history of blogging." We discuss "zines" and how that early paper-based product reflected the views of a pre-electronic community and resembled the electronic community in the way it spoke to its constituency. We finally get to the electronic options of computer-mediated communication by discussing audio and video blogs and the technology that supports these forms of communication.

The "Alleged" History of Blogging

Trying to identify the "real history of blogging" takes a leap of faith. Most longtime bloggers feel they know the real history of the practice. One only needs to Google "history of blogging" to find out how many people "really" know how blogging began. This section offers the best of what was available at the writing of this text.

Although no one wants to use Wikipedia as a source, most of us are tempted to at least check what it says on topics and find out what other sources agree with it. There is little to disagree with when Wikipedia points out that blogging became popular after the development of digital communities such as Usenet, GEnie, the CompuServe systems and early e-mail lists or Bulletin Board Systems (BBS).[48] This section includes the comments from supporting sources that address the early technology and approaches to computer-mediated communication (CMC).

WiseGeek identifies the use of the Internet forum as a discussion area on a website.[49] These websites allowed members to read comments from others and post their responses to them. The forums could be on any specific topic and the membership of those participating could increase or decrease in size based on the interest in the topic. The Internet forum was also called a message board, discussion group, bulletin board or Web forum.[50] While these communication options pre-date the blog, they had one big difference. The Internet forum allowed participants to respond to each other and change topics based on comments made. A blog, however, has one primary writer or a group of writers and others may only respond to comments made by the blog owner.

In late 1979, Usenet came into existence. Two Duke University graduate students decided to connect computer systems so they could share information about the UNIX operating system.[51] Not too far away, a University of North Carolina graduate student wrote the first newsreader software. The two universities were soon connected. The Usenet system grew throughout the 1980s. Other network administrators, volunteers and systems designers assisted the development of Usenet and gave it a backbone.

UNIX did not remain the sole topic of discussion by Usenet users. As more people began to use the system, the backbone had to create a system that allowed people with similar interests to converse more seamlessly. The backbone created seven top-level groups that included: computer discussion, miscellaneous topics, newsgroup issues, recreation and entertainment, science, social discussion and controversial topics. As expected, other topics were added because of their interest. Usenet created a system that allowed group members to suggest topics and vote on including them. Humanities was the eighth topic added.[52]

Although Usenet is an early participant in CMC's history, it still survives. Corporations like Microsoft, IBM, Novell and 3Com still use it. Also, the United States, Austria, South Africa and many other countries still use it. Corporations use Usenet now to educate their customers.

CompuServe Information Service (CIS/CompuServe) was an early contributor to the online community building efforts even though it is no longer a part of the community. In 1978, CompuServe began introducing a number of online services for personal computer users. They offered electronic mail, now known as e-mail, to help customers with service problems through its Infoplex.[53] It later pioneered "real-time chat" with its CD Simulator service in 1980. In 1981, CompuServe introduced a file transfer protocol that allowed clients to share files as well as allowed CIS to use the system internally. CompuServe was the first online service to offer Internet connectivity in 1989. In the early 1990s it was one of the most popular companies with hundreds of thousands of users participating in moderated forums.[54] CompuServe was able to move its use of CMC from the business client to the geek to the average computer user. Its online forums included discussions on entertainment, sports, politics and current events. In 1992, CompuServe hosted the industry's first electronic press kit for the film *Sneakers*.[55] CompuServe remained a leader in electronic forums until the mid-1990s. In 1997, AOL bought it out. It ceased to exist on June 30, 2009. Even in its death e-mail users can still use their system under a new system.[56]

GEnie stands for General Electric Network for Information Exchange. It was developed on October 1, 1985, by Bill Louden and launched as an ASCII text-based service later that month. This product was designed to compete with CompuServe's Community forum.[57] GEnie allowed users to communicate on GE's mainframe computer during hours when the computer was not used for business. It became very popular and a national force in the online market but it was never fully developed. GEnie developed a reputation for online text games as well as graphical games that used state-of-the-art non-textured 3D graphics on PCs with VGA displays.[58] General Electric, however, never saw GEnie as a product for profitability. It never added a phone line so more calls could come through and did not expand the mainframe to take more users. For some reason, General Electric was successful in reaching the masses, but did not try to keep them.

The use of Bulletin Board Systems (BBS) takes us closer to today's blogs. An examination of JSTOR research on early bulletin board systems suggests that the majority of the users were in education or seeking better teaching methods for specialized audiences or they reflected the needs of various professional groups. Like the other online communities, BBS reflect a single topic and allow all participants to respond to topics put on the table. In the 1980s, newspapers and broadcasters started dial-up bulletin boards.[59] At the time, these bulletin boards were run by engineering or newsroom geeks and they were often in partnership with America Online, Prodigy and CompuServe.[60] While other newspapers decided to work online, few made the commitment of the *Chicago Tribune.* In 1991, the Tribune Company invested in America Online. In 1992, the Tribune launched Chicago Online. Unlike the other mediated units, the Tribune marketing department drove this online bulletin board.[61] Founding editor Owen Youngman noted that content for that first bulletin board came from all over the paper during the first year. The next year the newspaper hired a staff of editors and turned the bulletin board into an online newspaper. Now within the newspaper structure, reporters have established their own blogs based on their reporting or commentary areas.

From the very basic goals of trying to reach the consumer, the precursors to the blog used every approach within their skill sets to come up with ways to get the attention of readers and viewers. This desire to attain and maintain attention did not begin with the advanced communication links.

"Zines"

Before blogs, there were "Zines." Zines is short for fanzines.[62] Zines are amateur publications created and distributed by one author or a team of authors. Zine articles are written by average people who have an obsessive interest in a topic and are willing to write about it.[63] According to Dykeman, zines have been published for many years and date back to Paul of Tarsus or Martin Luther. Dykeman notes that zines existed before the amateur press association (APA) began.[64] APA was started in 1876. The National Amateur Press Association is still functioning.

Zines and blogs are very similar. *The Inner Swine* identifies five key aspects of producing a zine compared to a blog.[65] *The Inner Swine* suggests that the creator must first decide what kind of zine they want to produce. Then the creator has to take the big step of actually writing the material they want to share and taking pictures that relate to the topic. Next, the zine creator has to typeset, lay out and collate the material for the zine. When all this has been done, the zine publisher has to distribute the material. The last stage is to do the first four steps over and over again. Ziners spent a lot of money printing their zines. They had to buy paper, photocopy the zines, then pay to mail them to those who might want them. Those who did not want to mail them would have to find locations that would distribute the zines for little or no cost. Locations used for distribution usually had some relationship to the topics discussed. Zine publishers also sought advertising when their zine became popular enough.

When reviewing the list, one can clearly see that the steps for producing a blog vary very little. Bloggers have to create hyperlinks, add their pictures and, if using the appropriate technology, add either audio and/or video to their blogs. Cost for blogging depends on the level of technology the blogger wants to use. Basic blogging costs no more than the cost of Internet service if the blogger pays for Internet service. Some bloggers use the public library. Bloggers can advertise by seeking the advertising support of related businesses. Bloggers can also get the name of their blog out through blogging reviews and directories.

Zines have another characteristic that carried over to the online communities. Zines were often anti-establishment. This fits the early online or virtual communities. People who agreed on a subject that might not reflect the general opinion of the populace agreed to share their views through zines and later through blogs.

What Is Blogging?

There may be more agreement about what blogging is "not" than about what it "is." According to Dykeman and a few hundred other bloggers, bloggers are not journalists. They may want to use journalistic principles but they do not need journalistic credentials.[66] In 2009 there were over 200,000,000 blog posts. More than half of the bloggers post daily on either their blog or Twitter sites. Bloggers select a topic and write about it. They then allow others to comment on their thoughts. Others are not able to change the topics of future blog posts. Their comments might encourage the blogger to continue the topic in future blog posts.

A blogger named Mike[67] noted that blogging is closer to broadcasting than to conversing. He asserted that blogging is a conversation of sorts. The conversation is more intimate than what we expect in traditional broadcasting. However, blogging does not have the same feel as personal human communication, in part because the response or feedback to comments made is delayed.

Audio Blogs

Blogging is clearly established in both print and broadcasting venues. Technology, however, has permitted all who blog to advance the interest and accessibility to their blogs. Audio blogging has added a new dimension to the blogging phenomenon. Ipadio is an application (app) for Android users.[68] Through Ipadio, users can stream live audio to their website or pre-record up to 60 minutes of audio clips for uploading or posting later. By talking into the Android phone, users can text, title, tag and produce a transcript for their blogs. The technology allows users to add up to four photos from their telephones. Geo-locating technology is also used in Ipadio. The app provides tools that allow the audio information to be added to Facebook and Twitter. SpinVox's voice technology is used to activate this audio service.[69]

Dr. Mark K. Smith, CEO of Ipadio says that since they launched Ipadio, the product has worked on all telephones: landlines, satellite, VOIP and mobile.[70] However, when they developed specific apps for the iPhone and Android phones, they began to offer additional functions for phones as well as for on-line accounts. Using Ipadio for blogs results in phlogs or phone blogs. These blogs appear on ipadio.com instantly or are indexed in just a few hours. As early as November 2008, Ipadio pioneered the live audio-to-web market.[71] Audio blogs are also used in government work. International blogging done by the U.S. government provides new technology and new information for international audiences. Radio Martí held Cuba's first ever blogger awards ceremony featuring the country's most popular bloggers.[72]

The United States government funds the Office of Cuban Broadcasting through the presidentially appointed Broadcast Board of Governors. The Board of Governors was established in 1990 to oversee the operations of Radio and TV Martí. Both broadcast units transmit news and information to the people of Cuba from a non-government controlled perspective.

Video Blogs

Just as audio has enhanced blogging, new video services were added in 2009. Another product designed to be an e-mail service also has a broadcast capability. HelloWorld[73] offers a "comprehensive web-based platform" that allows users to communicate by using text or video e-mail. The product also allows users to create, store, share and publish their digital assets. This product allows the users to perform in their own talk shows, music videos and other podcasts. Internationally distributed, HelloWorld allows users to respond to selected viewers, listeners and companies conducting contests.

Mobile Blogging

The discussion of blogging could not be complete without identifying some of the apps that utilize mobile technology. Falcon Mobile Video Blogging allows a subscriber to stream a video directly from his or her cell phone to any blog including to YouTube by simply calling a pre-defined number.[74] Phones that are used need to have at least a 3G system and mobile video blogging service. Falcon Mobile will allow users to publish their thoughts and feelings at any time and any place.

Leaving blogs to examine Twitter and Facebook is somewhat problematic. Just as blogging is easily associated with mobile Twitter, Twitter and Facebook are connected to blogging, and mobile. In the following section, the basics of Twitter and its very brief history are presented.

Twitter

Twitter began in March 2006.[75] By the end of 2009, it was expected to have accumulated 25 million users with revenues of $4 million. "Twitter is making deals to put its application on every cell phone, game console and mobile computing device in the world."[76] Twitter is quickly growing in popularity. Teachers are tweeting with their students and there even appears to be a bit of Twitter overload. Spreading like wildfire, Twitter has seemed to make a space for itself in our lives.

Based on *PC Magazine*'s online definition, Twitter is a website and service that lets users send short text messages up to 140 characters in length from their cell phones to a group of people. Twitter[77] was designed to keep family, friends and co-workers aware of their various daily activities. Once a user sets up a Twitter account, users can import their e-mail address books as well as use the Twitter search engine to locate and invite people.

Since its inception, Twitter has left the area of personal communication and moved on to commercial and organizational communication options. Advertisers use Twitter to wish customers happy birthday and let them know they can receive a free product. This approach is used routinely in South Korea by Starbucks on cell phones. Other businesses use Twitter to announce the arrival of new products. Politicians routinely use Twitter to inform their constituents or subscribers of their positions or opinions on events taking place in the city, state or country. Twitter is increasingly used for commercial organizations to tell customers "what's new." In addition, politicians and celebrities use it to keep constituents and fans informed.

Twitter messages ("tweets") are distributed to recipients who elect to become followers. Messages can also be sent via instant messaging, the Twitter website or a third-party Twitter application. A MySpace account can also be updated. To follow a Twitter feed, the Twitter site and feed name is the address (URL); for example, Best Buy's Twitter feed is www.twitter.com/bestbuy.

Twitterese and Users

As noted, a Twitter message is a "tweet," and an ongoing stream of Twitter messages is a "Twitter feed." People who write tweets are called "twitterers" or "tweeple," and a gathering of twitterers is a "tweetup." A novel written and distributed as tweets is a "twiller." A Twitter follower may also get stung by a Twitter phishing scam. Twitter's audience base is a younger demographic primarily comprised of 15-to-24-year-olds or the Millennial Generation. However, the playground is no longer secure. Not to be left behind, 25-to-34-year-olds have found their way in.[78] This use by an older group is driving the younger users away. With Facebook already populated by Baby Boomers and Twitter feeling the effects of Generation Xers, Twitterers are on the run for a place of their own.

The release of figures in 2009 indicated that the use of Twitter by 15-to-24 year-olds dropped from 55% at the beginning of 2008 to 50% in 2009.[79] While nothing suggests that 15-to-24 year olds are abandoning their Internet use, there is an indication that they are spending less time on social networking sites.[80] James Thickett, director of market research at Ofcm, said that youth under the age of 16 are remaining "immensely" loyal to Facebook and MySpace.[81]

Challenges to Twitter

As great as Twitter can be for getting micro messages across or recommending restaurants or businesses to others, Twitter's reputation has been compromised from time to time. In September of 2009, a series of micropostings swamped Twitter with malicious scams, thereby costing Twitter its reputation.[82] The scam was actually one that had been used in e-mail brought to the new platform. Dishonest Twitter users created Twitter accounts and sent tweets that carried links to promotions for fake antivirus protection that did not exist. "One wave keyed off Twitter's top 10 trending topics," spreading bad links in tweets purportedly about subjects generating the most microposts globally. The "direct messages carried links designed to steal passwords and recruit people for work-at-home mules, setting up bank accounts to help thieves extract funds from hijacked financial accounts."[83] One of the most disruptive effects of these acts of piracy was the loss of trust for Direct Messages (DMs). Sean-Paul Correll, researcher for Panda Security says, "People have a built-in trust in messages that appear to be from friends." Twitter's 30 million users make it a prime target for scammers and schemers. Therefore, use of known antivirus software that is frequently updated is a must. Achido[84] cautions Twitterers to use all the free tools available to them. Some can unravel shortened Web links before you open them and others let you know if the website is safe.[85]

Twitter and Broadcasting: An Expanding Relationship

There are clear indications that the broadcast industry does do business with Twitter. Besides creating television programming from Twitter-developed feeds, broadcasters are also using Twitter to promote programming. Technology entities are designing equipment that is adjusted to accept tweets and broadcasters are feeling the effect of unhappy viewers based on their tweeted messages to tweeple.

Twitter is proving to be an effective method of television promotion. A Media Post News Brief on November 10, 2009, identified a script ordered by CBS for a show that was launched as a Twitter feed. The show, written by Justin Halpern, is about Halpern's father's comments, which he seemed to hear more when he moved back home at 29. Max Mutchnick and David Kohan, creators of *Will and Grace*, served as executive producers of the CBS show called "Bleep My Dad Says."[86]

Fox Network is using Twitter on a few of its reruns. They will feature scrolling cast and director commentary.[87] Additionally, IBM's NYSE:IBM and Fox viewers will be able to communicate with their flat-screen TVs by using Facebook and Twitter.[88] Fox used two of its shows (*Fringe* and *Glee*) during the summer of 2009 to allow the viewers to talk to the cast members and producers, as well as to other viewers. The tweets could be followed on a computer or by scrolling messages at the bottom of the TV screen.[89] IBM has accepted tweeting as a viable part of their business operation. They have "patented a remote control that incorporates tweeting and other social media features into the television-viewing experience."

Facebook

Facebook was created in February 2004 by Mark Zuckerberg, who is now the CEO of the network.[90] In a November 2009 report, Facebook was said to have a 250 million user-base and more than $500 million in revenue.[91] "Facebook had more than 10,000 partners and is developing its own currency program."[92] A report given one month later showed the social network's user base as 350 million.[93] At this time, it is considered the world's number one Internet social network.

Facebook has shown its value in broadcasting. The presidential election of 2008 proved to be a motivator for the networks to decide to combine their brands. In 2008, ABC News and Facebook created a partnership to create a U.S. Politics application.[94] When the partnership began in 2007, Facebook had only 83 million users worldwide, 30 million in the United States and even fewer among the news organization websites.[95] In 2008, ABC had only a "miniscule percentage of Facebook's audience," but the network was interested in marketing on Facebook because it cost so little and if they failed to attract a massive audience, they would not lose a lot of money.[96] Additionally, if the traditional media used Facebook, it offered a high probability of attracting youthful masses to their brand.

MSNBC's Vice President for Marketing noted that adding a relationship with Facebook was a "low-risk opportunity." Catherine Captain commented that "some things hit and some miss,"[97] but broadcasters were hoping for the "word of mouth publicity because digital word of mouth snowballs on its own." That snowballing is sped up by selective media buys that place CNN, MSNBC and other news organizations' ads on members' profiles.[98] At the time the article was released, CNN, MSNBC and ABC were in the process of revamping their Web pages so that they would be more interesting to the younger audience base.

ABC was not the first of the big four networks to attempt to forge a relationship with Facebook. Fox News Channel has had a Facebook presence since 2006. In 2008, Fox had a fan base of more than twenty thousand. CBS News did not have a fan page at the time Fox and ABC were adding that option. It worked hard, however, to develop a Facebook presence. According to Wesprin, after the presidential primaries were over, ABC attempted to discontinue its use of the news app but instead had its brand attached to Facebook.[99]

Broadcasting and Cable called Facebook a ripe target for television news organizations because the site attracts an audience with an average age of 23. Facebook does well with the age groups of 18-34 and 25-52 according to CBM Advisors.[100] Facebook applications, like widgets that are embedded in user profiles, allow for greater interactivity. Facebook earns money through advertising. Facebook social ads can help garner more page viewers and grow the customer base.[101] While inclusion of ads has proven to be successful for some clients, Jennifer Leggio pointed out that too much advertising can upset the regular readers. TNT used Facebook to launch its Fall 2009 season to the point of overkill according to Leggio. Leggio and other blog readers agreed that saturation of ads on a website was "annoying."[102] Facebook's popularity has forced it to redraw its privacy boundaries.[103] The network has decided to open up part of its site to outsiders but they are introducing more options for users' privacy settings. This will give users more control over who sees their thoughts, videos, photos and personal information.

This effort to provide more privacy options becomes very important as Google and Microsoft continue to incorporate more user-generated content from social media websites into their search results. Also, as Twitter continues to compete with Facebook, giving the user more options seems important. Facebook wants to be able to allow its users to share information in a number of different ways. Users may want some information to be made available to their friends while other information might be made available to their family members. Elliot Schrage, VP of Facebook's Global Communication and Public Policy Office said that the changes the network will make will not change their policies governing the kind of user information shared with advertisers.

In mid-2009, Canada's privacy commissioner said that Facebook lacked active safeguards to prevent unauthorized access to users' personal information by third-party developers like game and quiz makers. Facebook began testing its public messaging feature in the summer of 2009. In October of 2009, Microsoft announced plans to incorporate Facebook messages flagged for the general public into its search engine results. Just before the end of 2009, Google announced plans to incorporate certain Facebook data in its real-time search product.[104]

Broadcasting and Facebook

Facebook is considered more of a broadcasting product than a social networking product. Dr. Cameron Marlow, the man identified as Facebook's in-house sociologist, believes this and offers support for why the statement is accurate.[105] Marlow provides statistics that support his beliefs. More specifically, *The Economist* found the following:[106]

The average *male* Facebook user with *120 friends*:
  - . . . comments on *7* of his friends' photos, status updates, or wall.
  - . . . messages or chats with just *4* of his friends
The average *female* Facebook user with *120 friends*:
  - . . . comments on *10* of her friends' photos, status updates, or wall.
  - . . . messages or chats with only *6* of her friends
The average *male* Facebook user with *500 friends*:
  - . . . comments on *17* of his friends' photos, status updates, or wall
  - . . . messages or chats with just *10* friends

The average *female* Facebook user with *500 friends*:
- ... comments on *26* friends' photos, status updates, or wall
- ... messages or chats with only *16* friends

So contrary to what one might expect, consumers are not "networking" as much as they are "broadcasting" their lives. Finally, the broadcasting industry uniting with Facebook will be here soon. Facebook broadcasting is listed in a search of the topic. By checking facebookbroadcast.com/[107] you will find that the site is set up but only holding the space. Perhaps that is the network's way of letting users know that inclusion of Facebook on Internet-based media is well on its way.

One can safely conclude that the broadcast industry has done well in its efforts to reach the masses. The masses continue to be reached on traditional television. The masses are reached on the Internet. Social media or Internet-based media is reaching millions. Just as the cell phone industry was used to connect countries that did not have an infrastructure useful to develop traditional phone service, social media has now given broadcasters access to audience members in almost any location in the world. Competition increases and technology changes, but reaching the masses increases all the more.

Notes

1. John Gross, "Social Networks, Texts Boost Fundraising," CNN.com, http://edition.cnn.com/2010/TECH/01/14/online.donations.haiti/index.html (12 February 2010).

2. Haiti Earthquake Twitter Updates: Live Real-time Pictures, HuffingtonPost.com, http://www.huffingtonpost.com/2010/01/13/haiti-earthquake-twitter_n_421722.html (21 January 2010).

3. Ben Parr, "Haiti Earthquake: Twitter Pictures Sweep Across the Web," Mashable: The Social Media Guide, http://mashable.com/2010/01/12/haiti-earthquake-pictures (21 January 2010).

4. "Celebrity Fundraiser to Help Raise $10 Million," ajc.com, January 17, 2010, http://blogs.ajc.com/haiti-earthquake-atlanta-responds/2010/01/17/celebrity-fund-raiser-to-help-raise-10-million/ (21 January 2010).

5. Christopher H. Sterling and John Michael Kittross, *Stay Tuned: A History of American Broadcasting, 3rd Ed.* (New York: Routledge, 2001).

6. Christopher Anderson, "National Broadcasting Company," *The Museum of Broadcast Communications*, http://www.museum.tv/eotvsection.php?entrycode=nationalbroa, (23 December 2010).

7. Anderson, "National Broadcasting Company," *The Museum of Broadcast Communications.*

8. Jack Gould, "New Radio Concept Would End Chains," jeff560.tripod.com, http://jeff560.tripod.com/tv5.html (Originally appeared in the *New York Times* August 10, 1945), (22 December 2009).

9. "Television History—The first 75 Years," http://www.tvhistory.tv/History%20of%20TV.htm (22 December 2009).

10. "Television History—The first 75 Years."

11. "Television History—The first 75 Years."

12. Elizabeth McLeod, "Early Commercial TV," *TV Broadcasting History-Various Articles*, http://jeff560.tripod.com/tv6.html (22 December 2009).

13. http://www.abc.go.com/shows/wife-swap (23 December 2009).

14. http://www.nbc.com/the-biggest-loser (23 December 2009).

15. http://abc.go.com/shows/dancing-with-the-stars (23 December 2009).

16. http://www.cbs.com/primetime/survivor/ (23 December 2009).

17. "Cable Television History," About.com, http://inventors.about.com/library/inventors/blcabletelevision.htm, (23December 2009).

18. http://www.fancast.com/ondemand (23 December 2009).

19. HBO On Demand, http://www.hboondemand.com (23 December 2009).

20. In Demand Networks, http://www.indemand.com (23 December 2009).

21. Original Interview Conducted by Maria Williams-Hawkins, held with Indianapolis Programming Directors, during the Fall of 2009.

22. Michael Cieply, "Writers Say Strike to Start Monday," *New York Times*, November 2, 2007, http://www.nytimes.com/2007/11/02/business/media/02cnd-hollywood.html (23 December 2009).

23. Nikki Finke, "Impact of WGA Strike on TV Production," *Deadline Hollywood*, November 14, 2007. http://www.deadline.com/hollywood/impact-of-wga-strike-on-tv-production/ (23 December 2009).

24. Tom Surette, "Law and Order Stars Return Sans Raises," *TV.com*, June 12, 2007. http://www.tv.com/law-and-order-criminal-intent/show/1381/story/9708.html (10 February 2010).

25. Randi Petrello, "Hawaii TV Stations Merge News Services," *Pacific Business News (Honolulu)*, August 17, 2009. http://pacific.bizjournals.com/pacific/stories/2009/08/17/daily24.html (10 February 2010).

26. Brian Stetler, "Advertising Losses Puts Squeeze on TV News," *New York Times*, May 10, 2009. http://www.nytimes.com/2009/05/11/business/media/11local.html (10 February 2010).

27. Wayne Friedman, "Americans Channel More TV: Viewing Up 20% From 1999," *MediaPost News*, November 10, 2009. http://www.mediapost.com/publications/index.cfm?fa=Articles.showArticle&art_aid=117161 (10 February 2010).

28. Wayne Friedman, "Child's Play: Kids' TV Viewing at 8-Year High," *Media-Post News*, October 26, 2009. http://www.mediapost.com/publications/?fa=Articles.printFriendly&art-aid=116102 (10 February 2010).

29. Friedman, "Child's Play: Kids' TV Viewing at 8-Year High," 2009.

30. Friedman, "Child's Play: Kids' TV Viewing at 8-Year High," 2009.

31. Friedman, "Child's Play: Kids' TV Viewing at 8-Year High," 2009.

32. Thomas Ruggerio, "Uses and Gratifications Theory in the 21st Century," *Mass Communication and Society*, 3, no.1(2000): 3-37.

33. Everett M. Rogers, *Diffusion of Innovations* (New York: Free Press, 1983).

34. http://www.aber.ac.uk/media/Documents/short/usegrat.html (10 February 2010).

35. John Smith, "Daily Newspaper and Television Use Drops in America," *Media News International,* September 24, 2009, http://www.mnilive.com/2009/09/daily-newspaper-and-television-use-drops-in%20%20%20%20%20America/comment-page-1/ (10 February 2010).

36. Smith, "Daily Newspaper and Television Use Drops in America," 2009.

37. Smith, "Daily Newspaper and Television Use Drops in America," 2009.

38. Television Bureau of Advertising, *TVB Online.com,* http://www.tvb.org/nav/build_frameset.asp?url=/multiplatform/Default.aspx (10 February 2010).

39. Search Engine Watch.com, SEM Glossary, http://www.searchenginewatch.com/define (10 February 2010).

40. Search Engine Watch.com.

41. Glossary, SPRYTHINK, http://www.sprythink.com/glossary.html (10 February 2010).

42. Glossary, http://www.robtex.com/dns/capilanou.ca.html (10 February 2010).

43. Bottle, "Glossary," Corking PR, http:// www.bottlepr.co.uk/glossary.html (10 February 2010).

44. Xeequa Corp, http://communitymanagers.pbworks.com/Glossary-and-Reference (10 February 2010).

45. Erik Qualman, "Statistics Show Social Media is Bigger Than We Think," August 11, 2009, http://socialnomics.net/2009/08/11/statistics-show-social-media-is-bigger-than-you-think/ (10 February 2010).

46. David Berkowitz, "Berkowitz List of 100 Ways to Measure Effect of Social Media," *Media Post,* November 17, 2009, http://74.125.47.132/search?q=cache:kveEe3diW1QJ:www.mediapost.com/publications/%3Ffa%3Darticles.showarticle%26art_aid%3D117581+Berkowitz+List+of+100+Ways+to+Measure+Effect+of+Social+Media&cd=1&hl=en&ct=clnk&gl= (10 February 2010).

47. Craig Calhoun, "Community Without Propinqity Revisited: Communications Technology and the Transformation of the Urban Public Sphere," *Sociological Inquiry,* 68 (3), August 1998, 373-397.

48. History of Blogging Timeline, http://en.wikipedia.org/wiki/History_of_blogging_timeline (10 February 2010).

49. Internet Forums, http://www.wisegeek.com/what-is-an-Internet-forum.htm, (30 December 2009).

50. Internet Forums.

51. History of Usenet, http://www.usenet.com/usent_history.html (30 December 2009).

52. History of Usenet.

53. Wiki:CompuServe, http://wapedia.mobi/en/CompuServe (10 February 2010).

54. Wiki:CompuServe.

55. Wiki:CompuServe.

56. Spalding Computer Services, http://www.spaldingcomputers.co.uk (30 December 2009).

57. GEnie, http://en.wikipedia.org/wiki/GEnie (30 December 2009).

58. GEnie.

59. Barb Palser, *American Journalism Review*, 10678654, Nov 2002, Vol. 24, no. 9.

60. Barb Palser, *American Journalism Review*.

61. Barb Palser, *American Journalism Review*.

62. Mark Dykeman, "The Secret Origin of Blogging That No One Discusses, *Broadcasting-Brain.com*, March 3, 2009, http://broadcasting-brain.com/2009/03/03/ secret-origin-blogging/ (29 December 2009).

63. Dykeman, "The Secret Origin of Blogging That No One Discusses."

64. Dykeman, "The Secret Origin of Blogging That No One Discusses."

65. Dykeman, "The Secret Origin of Blogging That No One Discusses."

66. Dykeman, "The Secret Origin of Blogging That No One Discusses."

67. Mike Sanders, "Blogging is More Broadcasting than Conversing," June 3, 2005. http://keeptrying.blogspot.com/2005/06/blogging-is-more-broadcasting-than .html (28 December 2009).

68. Android Guys, "Ipadio Releases Audio Blogging/Mobile Broadcasting App for Android," September 1, 2009, http://www.androidguys.com/2009/09/01/ipadio-releases-audio-blogging-mobile-broadcasting-app-for-android/ (28 December 2009).

69. Guys, "Ipadio Releases Audio Blogging/Mobile Broadcasting App for Android."

70. Guys, "Ipadio Releases Audio Blogging/Mobile Broadcasting App for Android."

71. Guys, "Ipadio Releases Audio Blogging/Mobile Broadcasting App for Android."

72. Broadcasting Board of Governors, "Radio Marti Covers Cuba's First-Ever Blogging Competition," September 10, 2009, http://www.bbg.gov/pressroom/press releases-article.cfm?articleID=429 (29 December 2009).

73. "Brand New Web-Based Video E-mail, Blogging Live Broadcasting," posted Las Vegas, Nevada, October 21, 2009, http://www.olx.com (29 December 2009).

74. "Innovative Connections Make a Difference by Windmobile," http://www .wind-mobile.com/?id=86 (29 December 2009).

75. Conrad Hall, "Social Media Marketing: It's The Television All Over Again," *Technorati*, November 2, 2009, http://technorati.com/business/advertising/article/ social-media-marketing-its-the-television/ (2 November 2009).

76. Conrad Hall, "Social Media Marketing: It's The Television All Over Again."

77. http://www.twitter.com (20 November 2009).

78. Richard Wray and Sam Jones, "It's SO Over: Cool Cyberkids Abandon Social Networking Sites," guardian.co.uk, August 6, 2009 http://www.guardian.co.uk/ media/2009/aug/06/young-abandon-social-networking-sites (30 December 2009).

79. Wray and Jones, "It's SO Over," guardian.co.uk.

80. Wray and Jones, "It's SO Over," guardian.co.uk.

81. Wray and Jones, "It's SO Over," guardian.co.uk.

82. Bryon Acohido, "Scammers Hit Twitter with Tainted Tweet Storm," *USA Today*, http://www.usatoday.com/money/media/2009-09-28-tweet-spam-twitter_N .htm (29 September 2009).

83. Acohido, "Scammers Hit Twitter with Tainted Tweet Storm," *USA Today*.

84. Acohido, "Scammers Hit Twitter with Tainted Tweet Storm," *USA Today*.

85. Acohido, "Scammers Hit Twitter with Tainted Tweet Storm," *USA Today*.

86. "Twitter Feed Becomes CBS 'Dad' Show, Media Daily News," *Media Post News*, November 10, 2009, http://www.mediapost.com/index.cfm?fa=Articles .showArticle&art_aid= (10 February 2010).

87. Tracy Swedlow, "Fox to Air Twitter—Enhanced Re-Run of Fringe and Glee," August 31, 2009, http://www.itvt.com/story/5527/fox-air-twitter-enhanced-re-runs-fringe-and-glee (10 February 2010).

88. Swedlow, "Fox To Air Twitter," 2009.

89. Swedlow, "Fox To Air Twitter," 2009.

90. "Facebook Redraws Privacy Boundaries," December 10, 2009, http://www.abc .net.au/news/stories/2009/12/10/2767063.htm (30 December 2009).

91. "Facebook Redraws Privacy Boundaries," 2009.

92. Conrad Hall, "Social Media Marketing: It's The Television All Over Again," *Technorati*, November 2, 2009.

93. "Facebook Redraws Privacy Boundaries," 2009.

94. Alex Wesprin, "New Divisions Retooling Facebook Apps," *Broadcasting & Cable*, August 17, 2008, http://www.broadcastingcable.com/article/115041-News_ Divisions_Retooling_Facebook_Apps.php (30 December 2009).

95. Wesprin, "New Divisions Retooling Facebook Apps," 2008.

96. Wesprin, "New Divisions Retooling Facebook Apps," 2008.

97. Wesprin, "New Divisions Retooling Facebook Apps," 2008.

98. Wesprin, "New Divisions Retooling Facebook Apps," 2008.

99. Wesprin, "New Divisions Retooling Facebook Apps," 2008.

100. Wesprin, "New Divisions Retooling Facebook Apps," 2008.

101. Jennifer Leggio, "Facebook Social Ad Overload," Turner Broadcasting, June 8, 2009, http://blogs.zdnet.com/feeds/?p=1274 (30 December 2009).

102. Leggio, "Facebook Social Ad Overload," 2009.

103. "Facebook Redraws Privacy Boundaries," 2009.

104. "Facebook Redraws Privacy Boundaries," 2009.

105. "Facebook Not a Social Network, but a Broadcasting Channel?" http://reface. me/news/facebook-not-a-social-network-but--broadcasting-channel/ (10 February 2010).

106. Justin Smith, "Facebook's In-House Sociologist Shares Stats on Users Social Behavior," *Inside Facebook*, February 27, 2009, http://www.insidefacebook.com/ 2009/02/27/facebooks-in-house-sociologist-shares-stats-on-users-social-behavior/ (30 December 2009).

107. Facebookbroadcast.com/.

11

Making Money with Mobile

Maria Williams-Hawkins

THE YEAR 2010 WAS AN IDEAL TIME TO publish a book on new media that allowed for the discussion of mobile technologies. The 2010 Consumer Electronics Show (CES), held each year in early January in Las Vegas, had numerous mobile media products to debut. Showcased were cell phones that were complete computers, cell phones that displayed live television shows, waterproof cell phones, and cell phones that were full projectors. The Consumer Electronics Show demonstrated that the conclusion of the twenty-first century's first decade and the transition to its second was one of rapidly changing mobile technology.

In this chapter, the current mobile market and especially its younger consumers will be examined. There will be a discussion of cell phone products that represent today's standards for mobile products as well as the latest improvements in mobile media. The challenges mobile companies are facing as they try to link Internet and other technological advancements to mobile technology are discussed. This chapter provides an overview of the history and current standards for mobile broadcasting. Finally, there will be a discussion of all the approaches advertisers and marketers are using to reach the masses via the mobile market.

Who Is Using Mobile?

The annual Consumer Electronics Show (CES) is one of the most anticipated events for technology enthusiasts. For vendors to invest so much time in the mobile arena, there obviously is a market for those products. Certainly cell phones are not new products, but they have changed considerably since their introduction to mainstream, middle-class consumers. From the luxury or business devices found in limousines, taxicabs and/or police vehicles to the micro-devices clipped to business users' ears, the cell phone has made dramatic changes since 1982 when the Federal Communications Commission (FCC) assigned an exclusive frequency range for cell phone use.[1] Mobile phone use continues to increase at an astonishing rate. Based on the Nielsen Company's report published by the Center for Media Research in January 2010,[2] there are 223 million cell phone users 31 years old and older. Moreover, many children 10 years old and older have cell phones but are not included in this figure. The number of mobile Web users has increased by 33% since 2008 with 60.7 million users. At least 18% of all cell phone users are smartphone owners. By the end of 2010, Nielsen estimates that 40% to 50% of all cell phone users will be smartphone users. Also, January 2010 statistics indicate that between 7% to 8% of mobile users have cell phones that stream audio and/or video as well as many other technologically savvy options. Not surprisingly, it is projected that smartphone use will increase to 150 million by 2011. Major questions remain such as which products will be the preferred products among consumers, which services will be most popular among consumers, and how communication companies will make profits in the mobile media marketplace.

The Youth Mobile Market

Despite a number of researchers focusing on the negative effects of cell phone use among children, including a Swedish report which suggested that children are five times more likely to run the risk of brain cancer[3] if they begin using cell phones during their childhood, more American families are finding early cell phone use to be beneficial when juggling busy family schedules. According to Mediamark Research Intelligence (MRI), a dominant voice in media consumer research in the United States, 20% of children between the ages of six and eleven own their own cell phones.[4] MRI further indicates that cell phone adoption by and for children has increased 68% in the past five years. Additionally, more than 12% of the parents interviewed say they plan to buy cell phones for their children.

Cell phones have been designed for children for more than five years. The Cingular Firefly is identified as the first phone designed specifically for children.[5] Limiting the features on the Firefly was seen as an adequate security measure that made the phone appealing to parents who wanted a safe means of cell phone use for their children. In 2005, the Firefly phones had no keyboards, picture taking devices or music playback. Currently, phones have those features, but are also equipped with additional security features to protect children. Some phones have GPS capabilities that allow parents to track their children's locations and children can take pictures of where they are and then send the photo back to their parents. Now, children's phones can do almost everything that phones marketed to adults are designed to do.

Security and convenience are key marketing factors for parents picking cell phones for their children. Parents want to be able to talk to their children when they want for the sake of convenience. The Marketing to Moms Coalition reported in the summer of 2009 that "mobile devices accounted for the top two ways mothers communicate with their children under 18 years of age."[6] Respondents indicated they talked to their children via cell phone 5.1 times per week and texted them 3.3 times per week.[7]

Interestingly, the two genders use cell phones differently. Over the past three years the number of boys who use cell phones increased by 50%. Girls increased their use of cell phones, but not to the same degree. According to Anne Kelly, "boys use more instant messaging, accessing the Internet and downloading games, music or streaming video."[8] Conversely, girls make more phone calls and send more text messages.

Besides the security features, cell phone companies can market their products to younger consumers for another reason. There may be an academic advantage for allowing young children to use cell phones. Studies show that children using cell phones are 36% more likely than other children to do their homework and are more likely to say that their parents let them go anywhere they want online. Additionally, parents believe that children who have their own cell phones are better multitaskers. This group of children text, talk, listen, and read simultaneously. The activities that they are less likely to do are download ringtones, send picture messages, or listen to music.

Hardware vs. Software

Mobile media users are in agreement about several things, including cell phone preferences. In a competition for customer satisfaction, the iPhone easily won.[9] In a study conducted by Claes Fornell International Group (CFI), based on a 100-point scale, the iPhone earns an 83 with the Palm Pre and T-Mobile's G-1 coming in second with a score of 77. Other contenders in the top five category include the BlackBerry with a satisfaction level of 73 and the Palm Treo with a satisfaction level of 70. Moreover, Verizon unveiled its own smartphone version called the Android. Later in the chapter, there will be an examination of the effect of the Android on the mobile phone market.

Doug Helmeich, program director for the CFI Group, noted that the iPhone "is the best thing to happen to the Smartphone industry because it captured the imagination of a whole new set of consumers that might not have made the Smartphone jump."[10] Indeed, according to AdMobnetwork, the iPhone OS worldwide market share grew from 33% to 40% in August of 2009.[11] "The iPhone and iPodTouch accounted for 27% of handsets used in the network." Nokia registered just below those two products at 24% while the N97 and 5800 XpressMusic touch screen devices came in fourth and fifth respectively.[12] In a July 6, 2009, posting, Walsh noted other factors related to the success of the iPhone. *Consumer Reports* pointed out that its new smartphone ratings reflect an increased emphasis on factors such as display, ease of navigation, multimedia, and messaging (at the expense of talk time due to poor battery life), and voice quality. The aim is to better mirror the growing importance of the non-voice use of Smartphones. These consumer preferences also suggest that the mobile user is now more focused on mobile as a total media experience and less as a voice communication tool.

Although the iPhone was the ratings leader for an extended period, competition with the iPhone existed before it was the ratings leader. While Apple was working on its phone, Google was also making efforts to enter the competition. In July of 2005, Google bought Android Inc.[13] The founders of Android Inc. formerly worked for other communication companies in design and development. The company had created software for mobile phones. When they joined Google's team they began to develop a mobile phone that was powered by the Linux kernal. The Android phones were created with a different approach than the iPhone. Much more collaboration took place to create this new approach to mobile phones. Android development was facilitated by the Open Handset Alliance (OHA), agreed to on November 5, 2007. The Open Handset Alliance, is a consortium of companies that include Broadcom Corporation, Google, HTC, Intel, LG, Marvell Technology Group, Motorola, Nvidia, Qualcomm, Samsung Electronics, Sprint Nextel, Texas Instruments, and T-Mobile. OHA's goal was to develop open standards for mobile devices.[14] The Android was the first product they created using this standard. Additionally, fourteen other companies joined the Android project on December 9, 2008, and included: ARM Holdings Plc, Atheros Communications, Asustek Computer Inc., Garmin Ltd, Softbank, Sony Ericsson, Toshiba Corp, and Vodafone Group Plc.[15]

The mobile Web landscape continues to change. On January 5, 2010, Google officially introduced its Android phone labeled the Nexus One. The introduction followed the leaks, reports and press releases that preceded its distribution of the devices to employees and media writers in December of 2009 and its actual unveiling in January of 2010.[16] The Nexus One has a 3.7 inch, Amoled display and a one-gighertz processor from Qualcomm.[17] Its 2.1 3D graphic display also has a trackball and large complement of technologies including a compass and light sensor. The Nexus One is very thin; it is about the thickness of a number two pencil. Additionally, the phone has a "5-megapixel still and video camera with a flash. The phone supports Google Maps, Facebook integration, and quick contact to switch between social and communication applications. Users can customize widgets and allow people to post images that move through its live wallpaper concept."[18] Because it was built using the open technology concept, other developers will be able to create apps for it, also.

Nexus One, as it advances, will pre-set its voice features. Google launched an application that allows users to search by voice. Nexus One provides a device that is voice-enabled. This device allows users to create e-mail or use Twitter by voice command. This feature is already available in Google Earth allowing users to search for a location by simply speaking the name of the place they want to locate.

With continued improvements to hardware and software, one might be led to believe that technology developers have everything under control. There are, however, concerns about some of the options users have on their cell phones. Making cell phones work like desktop and laptop computers is still a tremendous challenge. The next section of this chapter discusses how the industry is addressing those issues.

Challenges to Using Mobile for Search

Mobile, although gaining in popularity, still has much room for growth and preparation for the competitive world of entertainment. The first challenge for mobile devices was technological. A Nielsen Norman Group researcher tested mobile devices in a number of countries and across scores of sites and handsets. Mobile units designed for use in the United States were tested for usability improvements. Jakob Nielsen "found that usability may actually have degraded since he tested in 2000. Finding the local weather on a phone in 2000 took 164 seconds; in 2009, it took 247 seconds."[19] The researcher suggested that mobile users are still very "search-centric." The steps they take to get information follow a pattern. This approach may be very similar to what they once did on their desktop computers to get information. While those approaches may have worked well on old desktops, on mobile units where keys are smaller and space is compact, this approach increases the number of typing errors and frustration, therefore increasing the time it takes to get to certain sites. Although the iPhone had one user who found weather information in 18 seconds, the researcher concluded that convoluted steps to getting information contributed to mobile usability challenges. "If any additional evidence were needed for mobile-dedicated design's benefits, this example should surely suffice," says Nielsen.[20]

The metric from this report that was repeated in December 2009 throughout the media was 59%, the relatively low success rate of users pursuing tasks on their phones. The team equates those results with usability of the Web in 1994. "It was that bad," Nielsen asserts. The research pointed out that the success metric improved dramatically in direct proportion to the phone's power. While feature phones suffered an abysmal 38% success rate, smartphones leapt to a 55% success rate and touch phones had a 75% rate, which is almost on par with the 80% rates of the Web when accessed on a PC.[21]

For all of the hype around full-Web browsing on mobile devices, the survey found that the success rate among mobile sites averaged 64%, while the rate for full websites on handsets dropped to 53%.[22] Standard websites simply are not designed for easy interaction even on the better handsets. The researchers also found that general satisfaction was higher for mobile-specific designs than for full websites on mobile screens. "If mobile use is important to your Internet strategy, it's smart to build a dedicated mobile site," asserts Nielsen.[23]

The constant hardware and software advances in mobile media make deciding which is more important very challenging. As soon as technology developers improve the hardware, new software is designed to make the hardware more convenient. The marketing of both the hardware and software keeps all the companies in business. The next section discusses concerns of advertisers and marketers of cell phone technology.

Making Money with Mobile

Companies interested in marketing on mobile now have a bit more preparation for the game. The Interactive Advertising Bureau (IAB) released its first primer on buying mobile media.[24] This guide is directed toward advertisers and marketers. The IAB Mobile Buyer's Guide provides insight into the mobile ecosystem, mobile campaigns, key mobile ad terms, shortcode campaigns, and examples of various mobile ad executions.[25]

According to Jeremy Fain, IAB's vice president of industry services, "We designed this guide to provide everything an agency or marketer would need to create a mobile advertising campaign."[26] Further, Fain noted that ad agencies have advanced to multi-screen campaigns and now address mobile more frequently. IAB believes that the primer will help advertisers reach their customers better. "Marketers and agencies are now creating multi-screen campaigns and mobile is playing an increasingly significant role as more advertisers use it to reach their customers."[27] The new mobile guide was released in conjunction with the IAB's Mobile Marketplace, which featured several sessions focused on basics like how to make mobile buys, how mobile fits in with overall media budgets and how to know if mobile campaigns are working.

At a summer conference, IAB also featured "an emphasis on marketing on Smartphones," underscoring the ad focus on more sophisticated devices like the iPhone and BlackBerry because they generate higher levels of mobile Web use than typical mobile phones. Forrester Research has defined the Smartphone as "a mobile phone or connected handheld device that uses a high-level operating system, including iPhone OS, BlackBerry OS, Windows Mobile, PalmOS, WebOS, Symbian, and any flavor of Linux including Android."[28] Forrester further distinguishes smartphones from quick messaging devices (QMDs). QMDs have a Qwerty keyboard and/or a touchscreen, but run on a proprietary software platform rather than a smartphone OS.[29] Smartphones have also become key platforms for hosting an ever-increasing array of mobile applications, including branded apps and free ad-supported ones."[30]

Wirefly, a mobile comparison site, released a study on mobile comparisons indicating that "mobile customers' behavior doesn't always match their preference for phones loaded with features. For example, 64% purchased a device with full Web browser and e-mail, even though 46% don't use the mobile Web and only 15% identify it as a necessary feature."[31] Additionally, more than half (51%) of those who bought a device with video streaming capability haven't used it to watch video.[32]

Mobile media companies have been pleased to learn that "mobile data usage, including things like video, music and web surfing appears to be on the rise." During the first half of 2009, the major U.S. wireless carriers reported higher mobile data revenue. Telecom research firm Chetan Sharma reported that revenue from mobile data services during the first quarter had hit $10 billion in the United States; it was the first country to reach that level.[33]

Advertisers and marketers see an important future in the mobile media marketplace. An interactive marketing forecast released by Forrester Research projects mobile advertising will increase 70% to $391 million in 2010 to $1.3 billion by 2014. The firm called mobile one of the "most anticipated, least adopted" interactive channels in the mix. Discussions later in the chapter present technology advances that may prove these projections for advertising and marketing to be correct.

Strategies for Success on the Mobile Platform

As media businesses decide to offer programming for mobile users, they would do well to examine the challenges that others encountered when moving to Web-based programming. Smith[34] believes that the small screen size of cell phones will force publishers and developers to attend to the concerns or promises the technologies that preceded mobile streaming were supposed to have overcome. Smith presents a list of concerns that mobile publishers should continue. The list includes: core competency, clarity, targeting, and personalization.[35]

Mobile publishers have to attend to what is most important to their message. Screen size means that only the essential elements need to be shown. Many mobile publishers who entered early transferred their content to a mobile site with little thought about how the change in platform would affect those who would receive the content. Newsmax launched its mobile site in the spring of 2009.[36] The company learned that "the area that collects and links to conservative columnists around the web was among the most popular." Newsmax learned that collecting personalities was one of its core competencies. AgProfessional.com determined that presenting three main choices at the home page was most effective for them. Viewers are offered three lines and three links.

Clarity of content is essential. Clarity is related to both content and layout or design. Smith's 2009 study identified three brands that provide clear information on their websites that transferred effectively to mobile. Facebook, Weather.com and *USA Today* are said to have created great Web products but those same products work better on their mobile options.[37]

Targeting the audience and the appropriate advertising continues to be a challenge for mobile publishers. Too often, like early Web pages, mobile publishers put too many banners and performance-based add-ons on the screen at the same time. Additionally, mobile publishers must recognize that the small screen means that viewers can only attend to one message at a time. The location of information on the screen determines how much attention the user will give it. Publishers must target their main product to appear where it will get the most attention.

Personalization is one of Smith's greatest concerns. Social media is truly a user-controlled product. Consumers customize everything after they become acclimated to the product they have purchased. Mobile publishers give the users the ability to control what and often how and when they receive information they desire. In this regard, mobile is believed to be the platform that may cause the Web publishers to improve.

Mobile Advertising Exposure

According to Barr,[38] mobile Internet users are half as likely to click through banner ads when using their mobile devices than they are when they use their computers. Chitika, an online advertising network, analyzed 92 million impressions to assess click through rates. They found that mobile users had a click through rate of 0.48% whereas non-mobile users had a click-through rate of 0.84%.[39] Daniel Ruby, research director of Chitika does acknowledge that the commercials seen online are not the same commercials we see on television. Additionally, mobile users are less exposed to advertising online.[40]

Chitika's study revealed that of the 92 million impressions, 1.5% of the users were on mobile browsers.[41] More than 65% of the impressions came from iPhone users. Of the five major Smartphone operating systems, iPhone users had the lowest click-through rate.[42] Additionally, the majority of those engaged in serious Internet activities are iPhone users. Daniel Ruby pointed out the iPhone Safari system gives the truest browsing experience. He believes there are fewer mistake click-throughs compared to other systems.[43]

The mistakes that print-based companies made and overcame represent the beginning of layout concerns mobile media experienced. After overcoming layout issues for cell phones, service providers and producers had to learn how to give the consumer more of what they really wanted. Just as consumers wanted to carry their computers in their pockets, they want to have their televisions and DVDs with them as well. In the next section there is an examination of the efforts service providers went through and are still going through to get recorded, live and on-demand video to consumers' cell phones.

Broadcasting and Mobile

As social media permeate the technology available to us, it is impossible to expect mobile or cellular technology to be left out. Although research and trial products offering video for mobile dates back to the 1990s, use of video on mobile technology is still in its early stages. Broadcast, Internet and video industry representatives are still unsure of what will work best because there have been too few attempts to create consistent programming. At this writing, too few acceptable revenue streams had been developed. Only a few publishers have found profitable ways into the mobile industry. Those publishers include Bloomberg, *USA Today*, ESPN, Weather.com, and AP News.[44] Besides the publishers, other companies are still making attempts to establish their market in mobile broadcasting. This section examines the entrance of video and broadcasting into the mobile industry. The early efforts to provide video on mobile devices are discussed along with the developments as of the writing of this book. The section concludes with expectations of how long it will take for this part of the mobile industry to mature.

Early Efforts to Broadcast on Cellular

In the early part of 2000, the 3G networks could not deliver high volumes of live programming. By 2005, networks and content providers had a number of options for sending video to mobile devices. Service providers were unsure of whether to offer video on demand or live video or both. The early cellular service providers were limited due to the status of 3G technology in 2005.[45] Both Verizon and Sprint Nextel had early mobile video services using MediaFlo but Sprint did not offer the live service.[46] Qualcomm's MediaFlo is a "dedicated network that broadcasts video to all viewers at once like traditional broadcasting."[47] At that time, the biggest mobile service providers were: T-Mobile, Verizon V-Cast, VIDEO AGGREGRATOR, GoTV, MOBm, RTV (Real TV), and SMARTVIDEO. These providers' biggest concern was which format would make the most money.[48] Service providers had to decide on a format for transmission before determining what they could transmit. MobiTV and SmartVideo both offered live video that did not require re-editing or packaging the content. SmartVideo merged with College Sports Television (CSTV) and began to offer more sports programming and news.

In the first few years of 2000, a combination of American and European companies was used to develop America's approach to cellular video. Qualcomm's MediaFlo and Crown Castle's DVB-H were used to turn broadcast signals into mobile video. Verizon's V-Cast and Sprint TV used approaches that took broadcast signals, cleaned them up, and then sent the signals to the content providers, who encoded their files into the Windows Media Video format before sending them to Verizon and Sprint for transmission to their consumers. In 2004-2005, there were various broadcast and cable networks offering a number of options for mobile users.

By 2007, when the Consumer Electronics Show (CES) opened, Verizon Wireless and its parent company, Verizon Communication, presented a new mobile service. Verizon Wireless began to offer live broadcast TV on cell phones. Live broadcast service was set to begin by March 2007. The original broadcast was set to carry the television networks NBC, CBS, Fox and MTV after the programming was broadcast on their networks. The service was designed to broadcast both full-length and live shows.

Coincidentally, mobile users in South Korea had been receiving live TV on their cell phones since 2005. The United States was still acquainting itself with that technology in 2007. Mobile DTV saw mobile video as a way to maintain direct contact with viewers. FLO-T carried by AT&T, Verizon Wireless and Vodaphone PC charged a monthly fee of $9.99. FLO-TV users had to purchase specially modified cell phones to receive the best services. Its owner, Qualcomm, marketed a special device for receiving the broadcast service for $8.99 per month. Broadcasters believed that FLO-TV was too expensive. Currently, there are eight top service providers and five specialty carriers in the United States.[49] The top eight are Alltel Wireless, AT&T, Cellular One, Nextel, Sprint, T-Mobile, U.S. Cellular and Verizon Wireless. At the writing of this text, Verizon was identified as the provider with the greatest satisfaction level while AT&T was identified as the largest service provider

Educating Consumers

Brandon Burgess, chairman of the Broadcasters Coalition, discussed differences between what clients might want and what their organization felt they needed. Several companies and media articles promoted mobile users' access to simulcasts of network national programming. ESPN and other sports channels were always suggested as popular options. Burgess felt that before offering customers high-end services they needed to educate their customers.

Just as with any other industry, advertisers need proof of consumer interest in a product before they invest advertising dollars in it. Additionally, advertisers need to be able to measure the number of consumers' interest and know how they are interested in using a product.

In 2009, The Nielsen Company and Rentrak Corporation began working with the broadcasters' coalition. They were scheduled to identify an audience measurement system in January of 2010. Borrell predicted that Mobile DTV would not reach the $1 billion local advertising level until 2014. Borrell noted that "Radio didn't become people reading newspapers aloud and TV didn't become radio stars just standing in front of a camera. There's got to be a change in programming at some point."[50] Mobile programming has not come of age and identifying the appropriate advertising approach to reach that point will take more time.

Convergence Is in Sight

As mobile technology has advanced, trying to separate mobile from Internet has almost become impossible or at least very challenging. Based on Quantcast research, North America has experienced a 110% increase in Web traffic from mobile devices and globally there has been a 148% increase.[51] Although the statistics are impressive, one must remember that at this writing, Web-based mobile use is only 1.3% of overall Internet use; however, rapid improvement in operating systems and browsers is promoting rapid growth in this area.

At this writing, Apple is the dominant player and has the largest share of mobile page views.[52] Quantcast indicates that Apple continues to maintain 60% of the mobile page views. However, as indicated earlier in the chapter, the development of the Android phone is having an impact on the market. Motorola's Droid captured almost a 4% share the first month that it was available.[53] This has diminished the status of the Palm Pre and demonstrated that Google's Android operating system was likely to surpass the BlackBerry OS with a 12% share of the North American Web mobile traffic.[54] Quantcast anticipates that North American access to Web browsers through mobile devices should increase from 1.3% to 2.3% due to the development efforts of BlackBerry, HTC and Motorola.

Prospects for the Future

During the 1990s, prognosticators wrote about the day when viewers would be able to talk back to their televisions and be heard. During television's golden years, that was a commonly used joke and it became part of science fiction dramas like *The Twilight Zone*.[55] Early attempts were made to take us closer to that interactive level with WebTV and QubeTV. In the summer of 2009, a Texas based company filed a patent application for a remote control device that would let viewers blog, tweet or do Facebook updates without going to their computers. The remotes are designed to allow viewers to watch their TV shows and share their thoughts without leaving their seat. The remote simulates an old-fashioned practice of "talking to people."

Television continues to evolve. "A new category of internet connected televisions will include the CinemaNow service that allows users to download movies and other content through the TV."[56] Brien Steinberg projects that we are getting closer to the time when viewers will be able to access their online life with a TV remote control and/or big-screen TV will function more like a touch screen, knowing what we like to see and what music we want to hear. He believes that these televisions will even know what social media and e-mail information to send to us..[57] Viewers will get traditional broadcasting—on the flat screen or on their computer or their cell phones—and advertisers gain access to consumers. Thus, all are pleased.

As 2010 began, online publications touted the rise of all things mobile. Adam Fendelman updated his readers on new options for cell phone users.[58] At the Consumer Electronics Show (CES) the LG-produced Expo smartphone gained greater attention. The Expo uses Windows mobile and has a one-gigahertz processor. This serves as both a professional as well as a personal entertainment device in that it can project any presentations, photos or videos stored on the phone.[59]

The size of cell phones has presented many problems. An ongoing problem with the cell phone is its inability to function after getting wet followed closely by the problems many people have keeping up with them. In early 2010, T-Mobile is set to release the world's first waterproof cell phone, The Seal Shield, which is said to be completely waterproof. By spring of 2010, AT&T, Verizon Wireless and Sprint also planned to have the product out in their brand.

Forecasts for additional changes loomed for the year 2010 and beyond. In fact, 2010 was projected to be the year of the smartphone. The level of collaboration brought on greater possibilities. Many carriers offered Android devices and Charles Colvin, in one of his blogs, even noted AT&T began collaborating with the other companies. He believed that the cost of all Android devices would go down. With those predictions come the prediction that Microsoft and Nokia will see their market shares go down. The turn of the century saw a key set of leaders in mobile technology. The year 2010 may have marked the beginning of a new mobile era.

Notes

1. Korbin Newlyn, "History of Cell Phones—How This Omnipresent Device Progressed to Where It Is Today," http://EzineArticles.com/?expert=KorbinNewlyn (18 January 2010).

2. Jack Loechner, Center for Media Research, "Research Brief, Just the Facts," 6 January 2010, http://www.mediapost.com/publications/?fa=Articles.showArticle&art_aid=120071 (2 January 2010).

3. Geoffrey Lean, "Mobile Phone Use Raises Children's Risk of Brain Cancer Fivefold," *The Independent*, September 21, 2008, http://www.independent.co.uk/news/science/mobile-phone-use-raises-childrens-risk-of-brain-cancer-fivefold-937005.html (15 January 2010).

4. Anne Marie Kelly, "The Kids Are Alright: They Have Cell Phones," *Media Post's Engage: Kids 6-11*, January 14, 2010 http://www.mediapost.com/publications/?fa=Articles.showArticle&art_aid=120642 (14 January 2010).

5. Kelly, "The Kids Are Alright," 2010.

6. Kelly, "The Kids Are Alright," 2010.

7. Kelly, "The Kids Are Alright," 2010.

8. Kelly, "The Kids Are Alright," 2010.

9. Juliet Travis, "iPhone, Android and Pre Beat BlackBerry and Legacy Smartphones in CFI Group Customer Satisfaction Study," September 30, 2009 http://www.cfigroup.com/.../pressreleases/2009_Smartphone_pressrelease.doc.pdf (6 February 2010).

10. Travis, "iPhone, Android and Pre Beat BlackBerry and Legacy Smartphones in CFI Group Customer Satisfaction Study," 2009.

11. Rene Ritchie, "iPhone Now Owns 40% of Admob Network Usage," *TiPb*, October 1, 2009, http://www.tipb.com/2009/10/01/iphone-owns-40-admob-network-usage/ (8 February 2010).

12. Ritchie, "iPhone Now Owns 40% of Admob Network Usage," 2009.

13. Ben Elgin, "Google Buys Android for Its Mobile Arsenal," *BusinessWeek*, August 17, 2005, http://www.businessweek.com/technology/content/aug2005/tc20050817_0949_tc024.htm (15 January 2010).

14. Open Handset Alliance, "Industry Leaders Announce Open Platform for Mobile Devices," May 11, 2007 *Google Press Center* http://www.google.com/intl/en/press/pressrel/20071105_mobile_open.html (15 January 2010).

15. Jennifer Martinez, "More Mobile Phone Makers Back Google's Android," *Reuters*. Thomson Reuters, December 10, 2008 http://www.reuters.com/article/idUS-TRE4B86M120081210 (18 December 2008).

16. Laurie Sullivan, "Google Hails Nexus One as Convergence Device," *MediaPost News Online Media Daily*, January 5, 2010 http://www.mediapost.com/publications/?art_aid=120084&fa=Articles.showArticle (7 January 2010).

17. Sullivan, "Google Hails Nexus One as Convergence Device," 2010.

18. Sullivan, "Google Hails Nexus One as Convergence Device," 2010.

19. Steve Smith, "The 'Misery' Of the Mobile Web," *Mobile Insider*, July 21, 2009 http://www.mediapost.com/publications/index.cfm?fa=Articles.showArticle&art_aid=110220 (7 January 2010).

20. Smith, "The 'Misery' Of the Mobile Web," 2009.

21. Smith, "The 'Misery' Of the Mobile Web," 2009.

22. Smith, "The 'Misery' Of the Mobile Web," 2009 .

23. Smith, "The 'Misery' Of the Mobile Web," 2009.

24. Mark Walsh, "IAB Unveils Mobile Media Buying Guide," *Media Post News Online Media Daily*, July 13, 2009, http://www.mediapost.com/publications/?fa=Articles.showArticle&art_aid=109660 (20 November 2009).

25. Walsh, "IAB Unveils Mobile Media Buying Guide," 2009.

26. Walsh, "IAB Unveils Mobile Media Buying Guide," 2009.

27. Walsh, "IAB Unveils Mobile Media Buying Guide," 2009.

28. Mark Walsh, "Forrester: Smartphone U. S. Market Share Reaches 17%," *Media Post News Online Media Daily,* January 5, 2010, http://www.mediapost.com/publications/?fa=Articles.showArticle&art_aid=120085&nid=109540 (7 January 2010).

29. Walsh, "Forrester: Smartphone U. S. Market Share Reaches 17%," 2010.

30. Walsh, "Forrester: Smartphone U. S. Market Share Reaches 17%," 2010.

31. Walsh, "Forrester: Smartphone U. S. Market Share Reaches 17%," 2010.

32. Walsh, "Forrester: Smartphone U. S. Market Share Reaches 17%," 2010.

33. Chetan Sharma, "US Wireless Data Market Update-Q1 2009," http://www.chetansharma.com/usmarketupdateq109.htm (8 February 2010).

34. Steve Smith, "The 'Misery' of the Mobile Web," *Mobile Insider*, July 21, 2009 http://www.mediapost.com/publications/index.cfm?fa=Articles.showArticle&art_aid=110220 (7 January 2010).

35. Smith, "The 'Misery' of the Mobile Web," 2009.

36. Steve Smith, "Getting Mobile's Signal," *EContent*, May 2009, vol.32(4): 7.

37. Steve Smith, "The 'Misery' Of the Mobile Web," *Mobile Insider*, July 21, 2009 http://www.mediapost.com/publications/index.cfm?fa=Articles.showArticle&art_aid=110220 (7 January 2010).

38. Aaron Baar, "Study Finds Mobilists Less Likely To Click Through," *Media Post News Marketing Daily*, September 2009.

39. Baar, "Study Finds Mobilists Less Likely To Click Through," *Media Post News Marketing Daily*, September 2009.

40. Baar, "Study Finds Mobilists Less Likely To Click Through," *Media Post News Marketing Daily*, September 2009.

41. Baar, "Study Finds Mobilists Less Likely To Click Through," *Media Post News Marketing Daily*, September 2009.

42. Baar, "Study Finds Mobilists Less Likely To Click Through," *Media Post News Marketing Daily*, September 2009.

43. Baar, "Study Finds Mobilists Less Likely To Click Through," *Media Post News Marketing Daily*, September 2009.

44. Justin Smith, "Facebook's In-House Sociologist Shares Stats on Users Social Behavior," *Inside Facebook*, February 27, 2009, http://www.insidefacebook.com/2009/02/27/facebooks-in-house-sociologist-shares-stats-on-users-social-behavior/ (30 December 2009).

45. Marguerite Reardon, "Verizon Offers Live TV on Cell Phones," *ADNet News*, January 8, 2007, http://news.zdnet.com/2100-1035_22-150787.html (10 January 2010).

46. Reardon, "Verizon Offers Live TV on Cell Phones," *ADNet News*, January 8, 2007.

47. Reardon, "Verizon Offers Live TV on Cell Phones," *ADNet News*, January 8, 2007.

48. Ken Kerschbaum, "A Lot Goes Into a Little TV," *Broadcasting and Cable*, September 26, 2005, Vol. 135, Issue 39, p. 13.

49. Cell Phone Service Providers, http://www-phonedog.com/cell-phone--research/companies/default (18 January 2010).

50. Lauren Goode and Amy Schatz, "Mobile TV Gets Closer as Backers Cut a Path," WSJ.com, U.S. Edition, http://online.wsj.com/article/SB100014240527487034 2590457 (21 January 2010).

51. Mark Walsh, "Quantcast: Mobile Web Growing Fast, MediaPost News," *Online Media Daily*, January 5, 2005. http://www.mediapost.com/publications/?fa=Articles .showArticle&art_aid=120115 (8 February 2010).

52. Walsh, "Quantcast: Mobile Web Growing Fast, MediaPost News," *Online Media Daily*, January 5, 2005.

53. Walsh, "Quantcast: Mobile Web Growing Fast, MediaPost News," *Online Media Daily*, January 5, 2005.

54. Walsh, "Quantcast: Mobile Web Growing Fast, MediaPost News," *Online Media Daily*, January 5, 2005.

55. http://www.twilight-zone.tv/.

56. Brien Steinberg, "The Future of TV," *Advertising Age*, November 30, 2009. http://adage.com/mediaworks/article?article_id=140751 (22, December 2009).

57. Steinberg, "The Future of TV," *Advertising Age*, Novemer 30, 2009.

58. Adam Fendelman, "Cell Phones Blog," About.com, January 8, 2010. http://cellphones.about.com/b/2009/01/08/report-consumers-inc (20 January 2010).

59. Fendelman, "Cell Phones Blog," About.com, January 8, 2009.

12

Cinema in the Age of RWX Culture

Alexander Cohen

CINEMA IS UNDERGOING A TREMENDOUS SHIFT in how it is created, edited, distributed, and exhibited. While distribution technologies evolved almost continuously from the beginning of the medium's origin, the advent of digitization drastically altered these means and served as the condition of possibility for a vast change in scale inconceivable before the Internet. New distribution mechanisms necessitate new business models, but here at the beginning of the second decade of the second millennium, models remain by and large unsettled.

The low costs of independent filmmaking, the rapid unauthorized copying and distribution of cinematic products such as feature films by DVD and the Internet, and the increasing costs of traditional marketing and distribution have exerted such powerful pressures on studios that many say the system is in danger of breaking down completely. Some threats arise from media piracy from DVDs being uploaded to the Internet by individuals using personal computers. Others result from companies that have used certain "safe-harbor" provisions of the Digital Millennium Copyright Act (DMCA) to skirt the copyright issue long enough to allow tens of thousands, if not millions, of views before they are "taken down." Distributing films in whole or in part directly over the Internet has occurred through a variety of means from p2p (peer-to-peer) software and virtual network overlays on the Internet to various "private" upload sites and even e-mail as well as more recently by legitimate sites such as Hulu, YouTube, and Netflix. This was accomplished first through the Internet to personal computers and now, more recently, directly to personal video recorders (PVRs) such as TiVo and DirectTV's PVRs.

This chapter explores the impact of digitization and Internet-enabled distribution on cinema. While digitization acted as the condition of possibility for transforming cinema, it remains unable to account for both digitization and Internet-enabled distribution to cinema. Rather, the interconnection of personal computers via the packet-switching technology known as TCP/IP, alongside other technologies such as multi-tasking, multi-user computer-operating systems such as UNIX, combined with the domain-name system and the advent of the World Wide Web and search engines, proved to be the explosive combination. Indeed, it changed everything about media distribution in the first decade of the twenty-first century.

Film-industry executives were blindsided by the effect of the DVD and the Internet, although they already had much historical experience with media transformations. Everyone already had "seen this picture" during the origins of the VCR, in particular in the case of Sony's Betamax.[1] Before the VCR, the piano roll for the music industry and later the effect of Napster on the music industry fundamentally changed media distribution. Surely the pattern of change could be identified, and a general rule of thumb might be: What takes place in the realm of audio will eventually happen to the moving image.

The Internet is rapidly becoming the primary means of film distribution just as it is for music. Advances such as Blu-Ray will merely postpone but not eliminate the increasing adoption of Internet-based distribution and the concomitant "death" of hard media. Business models and technologies will have no choice but to adapt. Even if there remains a substantial number of consumers above age 25 who still enjoy the material satisfaction of having a box and a disk, the younger generation has little, if any, need for them. The concept of a show being broadcast at a specific hour on a specific day seems ludicrous to most 10-year-olds today. Similarly, the idea of going to a store to purchase a CD or rent a DVD is becoming equally arcane. As text has been nearly subsumed by the Internet and the power of search engines, so too goes the world of moving images. If that world is still fragmented by the sale of DVDs from Amazon, the rental of DVDs from Netflix, video on demand from the cable companies, or now the digital streaming of video to Personal Video Recorders, rest assured that it will not last long; business models are evolving and potentially converging rapidly.

There has been a gradual loss of control over time-based visual media such as cinema by virtue of the sheer number-crunching and networking power now in the hands of individuals with network-connected personal computers and "smart" cell phones. Similarly, there has been a colossal change in the way time-based visual media are consumed. When film was clearly in the hands of monopolistic film companies that controlled their production, distribution, and exhibition, we enjoyed what was primarily a "read-only" culture of film production. The scale and instantaneity of the Internet has led us from a "read-only" to a "read-write," and, finally, to borrow from the terminology of Unix file systems, to a "read-write-execute" (RWX) culture. Films and videos now are not only copied and distributed but also can be altered, transformed, and remixed in ways heretofore unthinkable.

An executable in Unix systems is a file that can be "run" by the computer, rather than just a passive file that is read or written to it as a program. It transforms a general computing machine into something more specific. In the context used here, "executable" is defined as a type of file with which the user can participate. Thus, rather than merely having a passive media object to "play," a media object is something a "user" puts to use. In no small sense the use-value of a media object has been drastically transformed in the last decade. Not merely passively accepted entertainment, media objects can be altered and transformed, used in a myriad of different ways, whether legally or not.

This evolution was concretized by Apple in the first decade of the twenty-first century when it created an ad campaign for its new iTunes digital juke-box application, and it at least temporarily legitimated what was formerly considered a mode of piracy: "Rip-Mix-Burn." Apple may no longer use that pithy statement, but by the end of that first decade it almost seems quaint, now that CD distribution is dying off and music downloads continue their assault on CDs. Such a rapid transformation has raised many questions: What is "film" in this digital era, and what is its relation with video? What does "live action" mean when animation, green screen, and other digital techniques have merged it so thoroughly that it is inconsequential whether something was "really" filmed or whether it was animated? What is the relationship between production and distribution for film, and how do the changes in both, by digitization and the Internet, lead to changes in exhibition and new forms of cinematic "execution"? Where do "home movies" and amateur filmmaking fit into the history of cinematic production, or more broadly into "screen practice,"? And how do the new capabilities afforded by digital video and Internet distribution affect the meaning of home movies and amateur production and their place in the history of screen practice?

Internet phenomena such as viral videos—the *South Park* short about 1996, and later, in 2003, the *Star Wars* Kid, various cinematic smashups and satires, YouTube personalities, and so on—all deserve to be understood as substantial phenomena and not mere aberrations. All, if not a part of cinema, are at least a form of "screen practice" that has existed since the early twentieth century.[2] There is not space here to consider all these questions, so this chapter will concentrate on trying to understand changes in the relationship between the author and the user brought on by changes in scale by distribution technologies.

Walter Benjamin's "The Work of Art in the Age of Mechanical Reproduction" argues that photography, and by extension cinema, can be inherently democratizing, even while its use can be corrupted by politicization. Digitization and mass distribution of cinema over the Internet not only make cinema more readily available than ever before, but, more important, in its digital form it provides the audience, its "users," with the ability not only to watch it (read-only) or copy it (write) but also to alter it for its own purposes (execute)—(i.e., also referred to as RWX). Thus, it is not only someone like Woody Allen, who can re-dub and create a work like *What's Up, Tiger Lily?*, but anyone can, in principle, with a relatively inexpensive computer and a borrowed connection to the Internet, download a film or rip it from a DVD, alter it, recombine it with others, critique, satirize, and remix it.

Digital Cinema

In principle, both video and film record moving images, but they have unique histories and modes of production, distribution, and exhibition. Chemically based film is a medium with a very direct causal relationship to what it reproduces. Aside from the lens and filters nothing intervenes between the light and silver halide crystals and subsequent pigments, which are developed. Film, even at its simplest, is a delayed medium that requires an additional apparatus for exhibition. Video, on the other hand, was originally a live broadcast medium but gradually, by virtue of tape recording, editing machines, and, most recently, digital-editing applications, has become indistinguishable from "film."

Reciprocally, film, once it could be effectively digitized, is now edited by the same video-editing software suites. With the development and standardization of High Definition Video and even higher resolution standards such as "4k," as used in the Red camera, chemically based film looks like it may eventually disappear as the recording medium of choice for cinema, just as it has for still cameras. There are differences between the two media still, aside from their phenomenological foundations. But these are aesthetic and are becoming less consequential every day as the "look" of film can be simulated. The vastly lower cost and flexibility of cinema production using digital cameras and computers means that film stock's days are surely numbered.

Further complicating the lexicography are new techniques afforded by digitization that include the seamless integration of live action and CGI (computer-generated imagery or really a form of computer animation) by green screen and other techniques as well as various forms of motion capture that allow an actor to animate a character by moving its body and facial expressions.

The terms "cinema" or "film" are used here to represent the general exhibition art form that one tends to associate with the word "film," and the term "video" is used to represent the medium for exhibiting shows such as serials like HBO's *The Sopranos* (even though it was shot originally on film). "Digital video" is used to refer to Internet-based forms of "video," including such forms as QuickTime, Windows Media, Flash, and Divx, among many others. "Digital Cinema" will be used to refer to the cinema that has undergone some form of digital transcoding and manipulation or, even further, using forms of motion capture to create what is really a new form of digital animation where actors can control their "avatars." Examples include the Coen Brothers' film *O Brother, Where Art Thou?*, which was one of the first films to use a digital intermediary for color manipulation in the former case, and James Cameron's *Avatar* for the latter.

Finally, there is the complicated case of animation, which introduces another set of lexicographic issues. At this point in the history of cinema, it would appear that animation as a technique is rapidly becoming a fundamental part of cinema. Motion capture makes dividing animation from "live action" nearly impossible. But this chapter does not attempt a full analysis of the ontology of cinema. For that, one would refer the reader to D. N. Rodowick's *The Virtual Life of Film,* which is considered the definitive analysis.[3]

One way we can distinguish between the exhibition forms presented on Television versus the Internet is by using the symmetrical terms "lean-back" and "lean-forward." These terms have been popular within the business community and effectively describe the way people tend to expect to view a time-based medium. "Lean-back" refers to the experience of watching films at a theater or video or films on video at home on the sofa. "Lean-forward" refers to Internet-based video and generally shorter-form entertainment viewed on a computer screen and where the consumer is also, and not inconsequentially, termed a "user." "Film," then, represents long-form cinematic narratives stored or viewed in any conceivable format, but this is the kind of cinematic work that requires the consumer to lean back and experience it. Typically, it is viewed as if it were intended in a continuous, uninterrupted viewing experience.

One also needs to consider how cinema, video, and digital video fit into the larger context of screen practice itself. To contextualize new digital works now distributed over the Internet, the concept of "screen practice," as enunciated by Musser and others, allows us to consider new digital video works within the history of screen practice in general. As Musser describes it in his book *The Emergence of Cinema,* some characterize the "living photographs" of cinema as nothing more than the next step in Magic Lantern exhibition.[4] Cinema was preceded by a form of image presentation that has been reborn in the digital age as the "PowerPoint deck." Indeed, individual images in a PowerPoint deck are called "slides," thus evoking their origins as part of the tradition of Magic Lantern exhibition. The explosion of new forms of moving-image content available on the Web on such venues as the Internet Archive and YouTube is the result of the ease with which digital video can be captured and edited, then uploaded, and, perhaps more important, distributed throughout the world nearly instantly. The scale of distribution available to anyone with a computer and Internet connection dwarfs any previous forms of distribution.

That scale has led to an explosion of new and various forms of recorded video cam sessions, trailer parodies, smashups, fan subs, machinima, video blogs, and YouTube videos. Almost none of these are cinema in any sense of the word, but they are part of a general history of screen practice. Just as the Edison Vitascopes were intended for the single viewer and were not to be shared, so too is the majority of digital video available on the Web intended for a lean-forward experience on either a personal computer or cell phone and generally watched in solitude, however widely distributed.[5]

The question of distribution has generally not been considered to be as important to the development, history, and continuation of the medium as its narrative artistic dimension, or at the very least it has not generally been well integrated into analyses of the medium. But the scale of distribution made possible by the Internet has had a fundamental impact on the way cinema and screen practice in general is currently developing.

The Question of Distribution: Audio Before Video?

In principle, the production and exhibition of cinema have always been linked inextricably to its distribution; once the question of how to create and exhibit moving images was solved, copying and distributing to its audience were fundamental to the economics of its production. But these issues were first played out in the field of audio recording. It should surprise no one that the social, cultural, and legal issues in the distributive mechanisms for audio have generally preceded those of the moving image. In the last decade the networked music-distribution program known as Napster transformed the audio industry in ways that are still to be worked out in the film industry. But the effects were similar: Napster provided the means for vast numbers of users to search, discover, and distribute music in ways that bypassed traditional legal and commercial distribution. Songs were freed from their connection to one another in albums and were traded freely across the Internet. It is probably the first display of a phenomenon unique to the digital world: global hyperabundance.

Material that was, in some cases, no longer in "print" was made instantly available. For probably the first time in history people obtained a glimpse of what the world could look like if everything were instantly available to anyone, anytime, anywhere. While clearly violating copyright, and in particular the newly minted Digital Millennium Copyright Act (DMCA), Napster blindsided the music industry. It took quite some time for the music industry to develop the legal tools necessary to shut it down. By then the damage to the industry had already been done. Rather than attempt to determine a way to work with Napster, which its founders and investors, in some form, understood was the only way forward, the music companies focused on destroying it. Apple Computer stepped into the fray, provided a form of digital-rights management that was largely invisible but acceptable to music distributors, and effectively took over music distribution. It has made a lot of money doing so. What eluded Napster and the music industry at the time was a business model, and it took Apple to figure it out.

But well before Napster, and virtually simultaneous with the development of the film industry, a set of legal battles took place in the world of audio recording and player pianos over copyright and new technologies that had developed. These were not covered by previous laws. The birth of the audio recording industry in the late 1890s and early 1900s and the copyright communications policy that came out of it prefigured and indeed became the model for copyright developments of the late twentieth century. And these, in turn, have had effects on business models and the development of technology and its evolution. To understand the social context in which VHS tape, cable TV, personal computers, and Internet distribution have developed, one must understand the legal issues and decisions that affected the cinema business industry and which placed limits in some cases on the uses to which these technologies may be put.

The record industry and the player piano both created problems for copyright that would become important to later Supreme Court rulings in the Betamax case with respect to personal uses of video recorders.[6] The Sony Betamax case was famously decided in favor of Sony, and while it did not help Sony against VHS becoming more popular and overtaking the Betamax format, the decision allowed people to make copies of broadcast programs, no matter what the content, freely for personal use.

The *Sony* case, as Pamela Samuelson has argued,[7] "is among the most significant IP decisions rendered by the [Supreme] Court during the three decades of Justice Stevens' tenure there because of its impact on the copyright and information-technology industries." The *Sony* case established the "safe-harbor" concept from copyright challenges with respect to non-infringing uses of technology.[8] Before *Sony,* and crucial to as well as directly cited by it, was the *Player Piano* case or, more properly, the *White-Smith Music Publishing Co. v. Apollo* case of 1908. Interestingly, neither of these new sound-reproducing technologies developed in the first decade of the twentieth century—one a means for recording the sound of a performance for later reproduction on a phonograph, the other a means for recording the performance itself for later re-creation on a player piano—paid any licensing fees to copyright owners for the songs they reproduced. Yet, by 1902, at least a million piano rolls had been sold, and, by 1899, 2.8 million records had been sold.[9]

These technologies introduced new ways for people to enjoy music. Audio recordings allowed time-shifting of live performances that challenged the status quo for the music business, then highly dependent on sheet music for individuals to play the music in their homes on their own. But copyright owners, Timothy Wu writes in the *Michigan Law Review,* "depicted the recording industry as irresponsible pirates whose reckless copying of music threatened to destroy American creativity."[10] As John Phillip Sousa told Congress in 1906:

> These talking machines are going to ruin the artistic development of music in this country. When I was a boy . . . in front of every house in the summer evenings, you would find young people together singing the songs of the day or old songs. Today you hear these infernal machines going night and day. . . . We will not have a [vocal cord] left. The vocal cord will be eliminated by a process of evolution, as was the tail of man when he came from the ape.[11]

The world that Sousa alludes to is, to refer to the introduction, a read-write-execute (RWX) culture, albeit not scaled globally as is known today. It was local and richer in face-to-face interaction but vastly more limited in terms of its distributive capacity by our current standards of communication. Behind the debates was the desire to protect the sheet-music business from a new technology in the form of piano rolls and "talking machines." Sousa worried that a culture would inevitably develop that was "read-only" oriented and would play back only recorded performances. Sheet music would no longer be "executable" by voice, which would atrophy to the point where no further music development would take place. But as the later settlement of the case would make clear, the issue was not actually about the future of creativity. Instead, it was about the business ecosystem: how to share profits among authors/composers, distributors, and the new technologies. This is what Tim Wu calls copyright's "communications regime," which is its management of competition among rival disseminators.[12]

Early cinema required methods of distribution and control for it to thrive economically. The technological, economic, and legal inventiveness necessary at the beginning of cinema cannot be overstated. There would be only thirteen years between the practical invention of the intermittent action Vitascope projector[13] that allowed for mass-audience film showings in 1895 to the formation of the Motion Pictures Patent Company (MPPC) that consolidated and standardized film stock, cameras, and projection devices with a patent pool. By that same year, 1908, there were some 6,000 movie theaters in the United States and more than twice that in another five years.[14] All manner of problems needed to be solved to create an industry that could manage to produce, distribute, and exhibit that much material.

By 1908 movies had already become a big business, and the commercial exploitation required the rationalization of film production and distribution. The MPPC consisted of patents assigned to it from the Edison Manufacturing Company, the American Mutoscope & Biograph Company, the Armat Motion Picture Company, and the Vitagraph Company of America.[15] The MPPC allowed for complex agreements among motion-picture manufacturers and importers, film exchanges, exhibitors, and Eastman Kodak. Film was distributed by a system of film exchanges that had to obtain licenses from the MPPC in order to function. Importantly only licensed exchanges could acquire films from licensed manufacturers, and film was no longer sold but rather leased.[16]

One often hears how rapidly the Internet and Web developed, yet film surely rivaled the speed of the commercial exploitation of the Internet in the late twentieth century. For example, the commercial World Wide Web effectively began in 1994 with the formation of the Mosaic Communications Corporation (thereafter Netscape Communications), and it introduced to an extra-academic audience the Web browser, which replaced text-based systems such as FTP (File Transfer Protocol) and Gopher. Napster was released just five years later in June 1999.[17] But it was not the Web browser but the release of Napster that proved to be the watershed moment for digital distribution on the Internet, although it lasted only three years before it was shut down by copyright-related litigation by the music industry.

By the time of the formation of the MPPC, the direction the film industry would head for the next ninety years was set and heading toward the formation of proprietary content and one form or another of near-monopoly control. This reached its pinnacle with the formation of the studio system in the 1940s. The Supreme Court ruled in 1948 to break up the studio system, which had integrated production, distribution, and exhibition. But even today there is a very tight relationship between the studios and the distributors, even if they have less control than they might have had at one time over exhibiting their films.

Until the beginning of Internet distribution of video in the first decade of the twenty-first century, outlets for film opened only gradually. The story of film distribution via cable TV and the VHS tape and the resistance by copyright holders to new technologies, as exemplified in the Sony Betamax case, is well documented.[18]

The repetition of certain historic themes relating to copyright and technology is not surprising, but it is quite fascinating. The fact that in audio and moving-image technologies no one initially paid any licensing fees to copyright owners of the songs they reproduced was repeated, in a striking manner, in the Napster case. None of the music distributed earned any money for those with distribution rights, let alone the recording artists themselves. It happened again about five years later with YouTube (now owned by Google), where individuals digitized or made digital copies of films and videos, among other things. Again, those with the rights to distribute video were completely bypassed. Unlike Napster, YouTube was more effective at using the safe-harbor protection provided by the *Digital Millennium Copyright Act* (DMCA). Sticking to the letter of the law, but not the spirit as an Internet host, YouTube would "take down" a video if it was properly informed, but it is a mystery as to how it was able to completely eliminate pornography from the website yet still be unable to find copyright violations by their users without assistance. In the case of copyrighted material, it was not in YouTube's best interest to do more than merely abide by the law, whereas with respect to pornography, YouTube might have been afraid that like other similar websites, it might be inundated with pornography to the detriment of their rather "clean" image.[19]

Another fascinating parallel is the argument that by putting up material for any user, Napster and YouTube appropriated others' creative content and threatened the economics of the existing copyright regime. For example, in the Viacom/YouTube *Complaint for Declaratory and Injunctive Relief and Damages*, filed in the U.S. District Court for the Southern District of New York, Viacom specifically argued that owners of the new technologies were making profits on material they did not own and had no rights to:

> Some entities, rather than taking the lawful path of building businesses that respect intellectual property rights on the Internet, have sought their fortunes by brazenly exploiting the infringing potential of digital technology.
> YouTube is one such entity. YouTube has harnessed technology to willfully infringe copyrights on a huge scale, depriving writers, composers, and performers of the rewards they are owed for effort and innovation, reducing the incentives of America's creative industries, and profiting from the illegal conduct of others as well. Using the leverage of the Internet, YouTube appropriates the value of creative content on a massive scale for YouTube's benefit without pay-

ment or license. YouTube's brazen disregard of the intellectual property laws fundamentally threatens not just plaintiffs, but the economic underpinnings of one of the most important sectors of the United States economy.[20]

This argument is, not surprisingly, remarkably similar to the *White-Smith Music Publishing v. Apollo Company* case discussed by Wu. That case was decided in favor of the Apollo Company, which manufactured player pianos and the piano rolls played on them. The ruling held that the piano rolls themselves were part of the machine and not simply copies of the copyrighted sheet music. Both the White-Smith and Betamax cases sided with the new technologies and thus made way for numerous other business and technological inventions. Inclusion of compulsory licenses for mechanical reproductions that did not require permission of copyright owners by the Copyright Act of 1909 to some degree ameliorated concerns of copyright holders.

Other similar recent cases include *Buma v. KaZaA* and *MGM v. Grokster*, both of which were on the same scale as Napster and YouTube with respect to distribution mechanisms. It is important, though, to distinguish between types of technologies at play here: Napster, KaZaA, and Grokster were all peer-to-peer technologies that have multiple uses. They can in principle facilitate file transfers between one user and another but are not broadcasting technologies in the conventional sense of one-to-many. P2P is inherently one-to-one. Any number of users might discover content on their sites, but in each case the file transfer is between peers. This is not the case with YouTube, and it explains why it will likely have a difficult time defending itself against copyright infringement.

Given the White-Smith and Betamax cases, it is perhaps surprising that Grokster and KaZaA were both shut down, despite being pure peer-to-peer networks. Napster maintained a master database, and thus it could be argued that it was at least technically feasible for them to perform take downs of files that were potentially infringing copyright. YouTube, however, is not a peer-to-peer technology; rather, it is a classic user-hosting website, and it has both a catalog of material and a search facility it fully controls. Arguments surrounding the Betamax video recorder are in some ways similar to arguments the entertainment industry made later in peer-to-peer cases relating to Grokster, KaZaA, and Napster. Like Betamax, these P2P technologies were largely, but not exclusively, used to copy content without consent by copyright owners. YouTube can make a similar argument in that a large amount of the content presented on its website is made by individuals for individuals. Yet, the main reason Grokster, KaZaA, Napster, and even YouTube grew so rapidly was clearly because of the copyrighted material that coexisted on these sites.

Betamax as a technology had an important difference compared to these sites because they were Internet-based and provided access to digital copies. The number of copies that could be made from an "original" was all but infinite. Whereas Betamax was limited by the cost of tape and the need to be physically distributed, no such limitations exist for digital copies over the Internet.

Time-Shifting and Fragmentation of Control

Sousa and many of his contemporaries were concerned that technologies that promoted passive experience over active involvement were a means for an evolution away from humanity as they knew it. If losing vocal cords is equivalent to losing the tail in an upright ape—that is, if the vocal cords are to become vestigial in an age when the performance itself is obsolete—then composers saw themselves as obsolete and destined to become extinct.

The ability to displace an actual musical performance to another time and another place shifts the control from the composer and the performer to a new category of human, the "user." The user is technologically endowed with the ability to play the music back at any time and any place at the expense, it is feared, of active or creative participation. Composers of that time did not imagine how recording artists might be the next type of composer.[21]

Sousa's concern was more about the money than anything else, for the recording industry had moved faster than copyright law. The composer would, in the absence of a enforceable copyright, be unable to reap financial rewards from recordings sales. Profits would instead go to the inventor and various businesses involved in the recording, copying, and distributing of the recorded sound.

Thus, copyright reform included composers who, unable to receive royalties on records or piano rolls, would have "no incentive to write or compose."[22] Among their many, sometimes overtly conspiratorial assertions, one argument has stood out that has sustained itself within today's copyright arguments: "The recording industry was actually helping composers by spurring the sales of sheet music; hence, no change to copyright was needed."[23]

This argument has played itself out repeatedly with the introduction of new technologies, first with the recording industry, then with radio and TV broadcasting, and now with the Internet. The arguments generally remain similar. The first, the one by Sousa, depicts new technologies destroying what is essentially if not quintessentially human (i.e., the human voice would prove to be a persistent theme). This theme was exhaustively analyzed some decades later by Walter Benjamin, whose essays "The Work of Art in the Age of Technical Reproducibility" and "On Some Motifs in Baudelaire" exhaustively analyze the effects of these modern technologies on the work of art and on the human sensorium. He came to a very different conclusion: The new technology of cinema in particular, its ability to shift time and space as well as the increasing fragmentation of experience, was not only democratizing but also constitutive of modernity.

Sousa's description of the effect the new technologies would have on cultural products, such as the visual arts and in particular music, was a response to the effects on art by mass production that appears to undermine any sense of cultural products as original or authentic and replace them with mobile and highly interchangeable simulations. But, as has been learned from Benjamin's essay "The Work of Art in the Age of Mechanical Reproduction," whatever one's sense of the loss of aura of the original performance or artwork, the mobility and interchangeability of the resource no longer mean replacing any pre-existing original, for the work of art has become the work of art "designed for reproduction."

Where is the original performance of a song created in the studio today by any recording artist? Where is the original of the film? They do not exist. The basic phenomenological questions these new technologies of reproduction of sound and of the moving image brought into play, as they traced the outlines of modernity, have been made even more complex. They are now interchangeable to the second order, super-distributed and transformed by the user, who sees them as remixable, as potentially something to be deconstructed, remixed, and reassembled.

It is this executable nature of the digital object, in conjunction with the near infinite distribution capabilities of the Internet, that leads to new emerging forms of screen practice by RWX (Read, Write, Execute) culture. Rather than restricting the technologies or the emergence of these forms, our culture needs to find adequate ways to balance the need for creators to be compensated with the need to develop possibilities of the new technologies. New technologies to balance those needs have yet to be developed, but their outlines are present within the arguments of the past.

Notes

1. The EFF maintains an extensive database of material on the Betamax case: See http://w2.eff.org/legal/cases/betamax/ (10 January 2010).

2. Charles Musser, *The Emergence of Cinema: The American Screen to 1907 (History of the American Cinema, Vol. 1)*, (Berkeley, CA: University of California Press, 1994), 15–54.

3. David Norman Rodowick, *The Virtual Life of Film* (Cambridge, MA: Harvard University Press, 2007).

4. Musser, *The Emergence of Cinema,* 15.

5. Solitude has contributed to the vast expansion of pornography on the Web, something that has been a part of cinema since at least the first peep shows.

6. Timothy Wu, "Copyright's Communication Policy," *Michigan Law Review,* Vol. 103, No. 2 (Nov. 2004): 297.

7. Pamela Samuelson, "The Generativity of *Sony v. Universal:* The Intellectual Property Legacy of Justice Stevens." *Fordham Law Review* 74, 1831 (2006): 101–145.

8. Samuelson, "The Generativity of *Sony v. Universal,*" 101.

9. Wu, "Copyright's Communication Policy," 298. See also Wu's source. See White-Smith, 209 U.S. at 16-18; Litman, supra note 55, at 350 n.70 http://supreme.justia.com/us/209/1/case.html (9 November 2009).

10. Wu, "Copyright's Communication Policy," 298.

11. Wu, "Copyright's Communication Policy," 278.

12. Wu, "Copyright's Communication Policy," 288.

13. The Vitascope was not actually invented by Thomas Edison but rather by Francis Jenkins and Thomas Armat. See Musser, Charles, *Before the Nickelodeon* (Berkeley, CA: University of California Press, 1991), 40-41.

14. Frederick Wasser, *Vini, Vidi, Video: The Hollywood Empire and the VCR.* (Austin, TX: University of Texas Press, 2002), 24–25.

15. Musser, *Before the Nickelodeon,* 304.

16. Musser, *Before the Nickelodeon,* 442.

17. John Alderman, *Sonic Boom: Napster, MP3, and the New Pioneers of Music* (New York: Basic Books, 2001), 103.

18. Jane C. Ginsberg, "Copyright and Control Over the New Technologies of Dissemination," *Columbia Law Review,* Vol. 101, No 7 (Nov. 2001): 1613–1647.

19. In the interests of full disclosure, this author (Alexander Cohen) ran a non-profit organization called Undergroundfilm.org, which focused on providing a legal outlet for independent filmmakers on the web. Undergroundfilm.org had a strict policy against using any copyrighted works for which the filmmaker did not have rights, which included music, images, and clips. The legal counsel was of the opinion that the safe-harbor clause of the DMCA would not protect websites that had any measure of control over what was made available on the website to users. Started in 1999 by Mike Kelly, this author took it over and while managing it, converted it into a 501c3 in 2003 after Pop.com, a joint startup by DreamWorks and Imagine Entertainment, which had purchased it, decided to close the entire project days before Pop.com went live on the Web. Cohen was Pop.com's CTO.

20. Viacom International Inc., Comedy Partners, Country Music Television, Inc., Paramount Pictures Corporation, and Black Entertainment Television LLC v. YouTube, Inc., YouTube, LLC, and Google, http://w2.eff.org/legal/cases/viacom_v_google/ViacomYouTubeComplaint3-12-07.pdf (10 November 2009.

21. It took a while, but because Bing Crosby no longer wanted to perform live on a regularly scheduled radio show, he invested in a company called Ampex which

legitimized the use of tape recordings for radio networks in 1948. See Mark Clark, "Suppressing Innovation: Bell Laboratories and Magnetic Recording," *Technology and Culture*, Vol. 34, No. 3 (Jul., 1993): 533 footnote 71.

22. Wu, "Copyright's Communication Policy," 300–301.

23. Wu, "Copyright's Communication Policy," 301.

13

Local Market Radio

Programming and Operations in a New Media World

Tony R. DeMars

I T MAY BE ALMOST UNIVERSALLY ACCEPTED that digital media is having and will
continue to have a significant impact on traditional (also called legacy)
media.[1] The model of how information and entertainment reaches an audience is in a state of change, and local market radio broadcasting is being affected by the transition. Anyone under the age of 30 today likely has had no
experience with free over-the-air television, having grown up with cable and
satellite TV. Younger consumers are also mostly disconnected from the idea
of using a VCR, a device that a generation ago revolutionized how the audience used television. This younger audience in a similar way may also have
little if any connection to local market radio stations the way previous recent
generations have.

The linear model of electronic media content distribution is increasingly
being displaced by new technologies and what might be described as a non-
linear or multi-linear model. While some investigators[2] found little research
related to these changes as elements of digital media use or about time spent
with various media, others suggests a recently emerging diverse range of empirical and theoretical research offers new opportunities for understanding
the impact of digital media, especially the Internet, on social, cultural and
socialization influences on the lives of youths and young adults.[3]

An Internet connection has become an important component of a contemporary American household, and the youth of the home play an important
role in adoption and use of the Internet.[4] The mainstays of traditional broadcasting—music radio and television programs—are still important components of the media mix of modern youth, yet media and communication

researchers and critics generally agree that a sea change is occurring in the media uses and habits of young people.[5] Industry data show almost 95% of U.S. Internet households had broadband connections by early 2009 and that the overall U. S. broadband penetration level was 63%.[6] Access to information in new ways means differences are inevitable in how users get content.

Challenges for Traditional Media

As the relationship between media content and the user continues to change, traditional media managers try to survive. Today's local market radio station managers wonder if the iPod, podcasts, music downloads and Wi-Fi radio have the potential to kill traditional local market radio broadcasting.[7] News convergence efforts that incorporated the Internet while coordinating with the traditional media has been one attempted survival technique for the television news industry.[8] In a similar way, entertainment creators are trying podcast clips for Internet or mobile listening or viewing, sometimes connected to the traditional media form they make and distribute by traditional means.[9] As an even deeper threat to the advertiser-supported media model, advertisers increasingly use digital media within their own development and implementation, that is, they are doing digital media content distribution on their own, rather than simply moving their advertising buys from traditional media platforms owned by others over to new media platforms owned by others.[10]

This chapter provides an overview of local market radio broadcasting framed within these realities of how traditional media have responded and are responding to the new competitive environment created by new audio platforms related to the explosion of growth of the Internet. This chapter explains and discusses economic and management issues related to this change and attempts to describe what local market radio managers believe is wrong with local market radio and what needs to happen for traditional radio broadcasting to survive. Programming and profitability aspects of local radio operations are revealed as a guide to understanding and managing local market radio in today's media environment.

Economic Implications: The Technology Perspective

In some ways, basic economic factors may have played an important role in current local market radio station problems. Things got too easy to do cheaply. Supported by relaxation of federal ownership limits, big group owners started running streamlined operations that maximized profits, but at the same time made local radio stations sound like canned, assembly-line copies of everything else. This level of voice-tracking and automation became most associated with the largest group owner, Clear Channel, and also came to be called "McRadio" in reference to being the media equivalent of having a McDonald's restaurant in every town. The size of Clear Channel became possible because of changes in the 1996 Telecommunications Act, but then competitors accused them of taking advantage of relaxed ownership rules beyond legal means and exerting monopoly control in local markets.[11]

There are two major issues with the economics of running a local market radio station after personnel: technology and competition. Early radio technologies required more work and required more people to put programming on the air. While this chapter is not intended to provide full details about early radio technology, a basic understanding of where we have been is important to understanding where we are.[12]

In the early days of radio broadcasting, as it emerged in the 1920s, many programs, and in particular advertising messages, mostly had to be presented live. Radio's first ability to record and play back content came through electrical transcriptions.[13] With the introduction of magnetic tape recording in the 1940s, commercial and program content increasingly was pre-recorded, and by the 1940s and 1950s playing music from records was commonplace and practical. By the 1960s, cart machines[14] had made playback of commercials and other short program content easy to broadcast.[15] However, despite the ability to do an automation system once carts and audiotape reels could be mechanically synchronized, inserting announcer voice segments to simulate a live sound was not practical, and the system was not nearly as efficient as a computer-automation system in various other ways. Federal Communications Commission (FCC) regulations also prevented radio stations at this time from running unattended.

What might be referred to as the late analog days, a radio station's operation was mostly done with a live board operator and announcer (typically with one person doing both, but in some situations with one technical person and one announcer). The technical process of automation involved cart carousels and open audiotape reels, and network services like newscasts came through phone line connections that had a noisy sound quality. By the 1970s and 1980s, radio network programming switched from phone lines to satellite distribution, and music-based radio formats became feasible, since the audio from satellite-distributed programming was broadcast quality. Satellite Music Network thus began in 1981. Satellite music networks originally were also limited by mechanical switching between the satellite feed and the local broadcast station's inserted commercials, promos and production elements, but could be made to sound reasonably local if implemented correctly by the local station. On-air sound quality continued to improve as stations switched over in the 1980s and 1990s to playing CDs instead of vinyl records. Into the 1990s, many automation systems still used a computer interface operating a carousel of CDs, but in the past ten years automation has become based on a completely self-contained software program, like Music Master.[16] There are several popular automation systems and some programs are even available as freeware. Completely self-contained programs can create playlists by randomly selecting music within categories, can insert all commercials and all other non-music production content including announcer voice tracks, and can seamlessly present the entire station's audio to the listener.

What's bad about the technology as it is presented today? An announcer sits in a studio and records his or her voice for multiple stations, sometimes with all those stations being in one city, but often for locations across the country, for the slots within all the program schedule where the announcer would talk, which is then mixed in with the music, commercials, jingles and so forth—essentially an assembly-line approach to creating a radio station's programming—and which can then be run by the computer as a kind of virtual programming running on autopilot.

Into the new century, as the speed of computers and related sophistication of automation software has improved, stations have the ability to network computers at facilities across the country, and the ability to transfer digital files easily through the Internet. Where in the past pre-produced syndicated programs might have been copied onto vinyl (and then later CD) and mailed to stations, now these can all be placed on a server and downloaded by stations into automation systems. The satellite delivery to local stations includes the 24/7 radio networks, some that actually voice-track and automate the programming fed to stations instead of having live announcers, but also allows for live syndicated shows for dayparts like *Delilah After Dark* as a nighttime show[17] (discussed more later) and the *Tom Joyner Morning Show* for mornings.[18] Even individual programming like a newscast can be easily created within a software program with a computer, such as news (as noted in the KNDE news example later in the chapter).

To top all these issues off, the old days of open reel tape recording, electrical transcriptions, rotary pots and tube transmitters have been replaced by technology that no longer even requires a transmitter. In the coming years, Internet-only "radio stations" will find that Wi-Fi radio will allow listeners to find them anywhere an Internet feed is available, and that will be virtually everywhere.

Economic Implications: The Competition Perspective

How much gross revenue should a radio station in one of the largest markets expect annually? If commercials in the largest markets average $300 each, and the station can average 80% sell-out based on twelve spots per hour from 6 a.m. to midnight Monday through Friday, gross revenues get into the range of $15 million. In the best days of local radio economics a few decades back, the top billing radio stations in the largest markets could gross about $30 million.[19] First Research notes that the 50 largest radio companies, which include Clear Channel Communications, Cumulus Media, Citadel Broadcasting, and CBS Radio, account for about 75% of the total revenue in the radio industry nationwide and the top four largest of those account for nearly half. Based on recent data, an individual medium-sized radio station had annual revenue of around $5 million.[20] First Research's publicly available proprietary data also shows the importance of these figures: the large companies have such a market dominance in individual markets that small companies have difficulty competing for advertising dollars.

Recent radio station sales show current local market radio station values and gross revenues. One enlightening example is the September 2009 sale of WCRB-FM, Boston, market rank 10, from Nassau Broadcasting to WGBH Educational Foundation for $14 million. In comparison, the original WCRB in the market (a different station with different technical qualities, and now WKLB) sold in the Boston market in 2006 for $100 million. Other recent radio station sales reported by George Reed also illuminate economic challenges to the industry. In a sale announced December 2008, KIMN-FM, KWLI-FM (now KWOF-FM), and KXKL-FM in Denver were sold by CBS Radio to Wilks Broadcast Group for $19.5 million. With net revenue of $12.7 million, a broadcast cash flow margin of 40% and broadcast cash flow at just over $5 million, the sale was 3.9 times cash flow and based on gross revenue of $14.4 million. Broadcast stations in the heyday of radio station profits were typically selling for seven to twelve times cash flow.[21] In July 2009, in San Francisco, the fourth largest radio market, Royce International bought KNGY-FM (now KREV-FM) from Flying Bear for $6.5 million. The station had been purchased by Flying Bear for $33.7 million. George Reed says the current market in buying and selling radio stations indicates a consistent pricing story—sellers are losing money and buyers are getting bargains. Only time will tell if these much-lower prices are indeed bargains, as values and profitability may continue to fall.

Radio station trading has been extremely low while selling prices have plummeted. Reed says further that no senior debt is available for transactions, and improvement in selling prices cannot come unless that changes.

Further, if one or more of the major group owner companies fails, there will be an ever greater glut of stations available for sale, with forced asset sales. A few of the positives noted by Reed are improvements in the values of public broadcasting stocks and the availability of private equity funding from outside traditional sources.

So is digital media competition killing local market broadcast radio? Mark Ramsey commented on the Fall 2009 Nielsen "Council for Research Excellence" study which observed several hundred people over two days to determine (among other things) what media they consumed and how they consumed those media.[22] This study revealed that, despite public perception to the contrary, radio continues to dominate as an audio medium, including among 18-34-year-olds. The assumption is that, with radio universally available and easy to use, it is still the perfect medium for passive listening. Ramsey warns, however, that as the distance narrows between ubiquity and ease of use, and with a current ambivalence toward local market radio in general, digital media has the potential to continue eroding the traditional radio audience.

What Radio Management Thinks

Discussion with local radio staff can help us understand how operations work in today's broadcast radio station. Chace Murphy, the News Director at Bryan Broadcasting in Bryan–College Station, Texas, notes that less than 10 years ago, he worked with paper scripts, the transition was still taking place for news "wire copy" from the Associated Press being searched on a computer and printed out to read, and audio was still often recorded on tape. In 2009, his newsroom is paperless and tapeless. Audio recording has progressed from analog audio tape to MiniDisc and flash-based audio recording, but now his audio from outside sources is based on real-time downloading and immediate file transfers. He sits at one computer station using the Wire Ready software program[23] to collect audio files from networks like the Texas State Network and ABC Radio, create an outline of a newscast, and edit the audio files and text of his newscast. He then goes to another computer connected to the audio console and implements the newscast completely from computer—reading the script and playing the audio files all from within the program.

On the music programming side at Bryan Broadcasting, KNDE Program Director Tucker Young says the Web "has become the new station van." While the station still goes out and connects with the public with promotional activities, promotions must include the Internet. Like many stations across the country, KNDE uses loyalty points on the station's website to engage listeners, has a Facebook group and uses the website as a component of the station's contests.

Bryan Broadcasting's Chief Engineer Chris Dusterhoff points to how technology made the work easier, but then made a lot more work for each staff member to do. Dusterhoff says the station has been doing Internet distribution of their stations since 1996, but sees an explosion of activity and growth emerging now. What are some economic challenges? The station had to pay an extra fee to run the *Rush Limbaugh Show* on their talk station's stream, but then had good listenership from people in their local market to the show from their Internet feed versus their broadcast delivery of the program. Noticing that most of their stations' Internet listening is 8:00 to 5:00 Monday through Friday, the Internet feed seems to make it easy for people to listen to "the radio" at work via the Internet. Bryan Broadcasting uses data from Google Dashboard to track where people listen to their streams and which programming does best. Being in the local market of Texas A&M University and having the contract to carry Aggie football games, so far they find their listenership remains fairly even through all days and times, but gets a huge bump in visitors when an Aggie football game is on. Bryan Broadcasting is one of those local groups of stations that seems to be using digital media technologies in the most practical way and, with such an approach, can easily deliver programming to an audience by traditional means or by digital distribution. Herein lies the problem.

According to Dusterhoff, the biggest economic challenge facing local market radio stations is the SoundExchange fee. The negotiated settlement between SoundExchange and the National Association of Broadcasters has stations paying a per-performance rate of $.0015 in 2009, $.0016 in 2010, $.0017 in 2011, $.0020 in 2012, $.0022 in 2013, $.0023 in 2014, and $.0025 in 2015. These are the per-performance rates that radio stations pay to Sound-Exchange.[24] Radio stations must file a complete accounting of every song that played and how many people listened to each song. Each listener of each song is considered a performance. Dusterhoff notes you can approximate the budget it will cost each station by assuming 12 songs an hour times the aggregate tuning hours from previous months plus a growth rate. In this example, in a month taken from Fall 2009, KNDE Radio had 18,859 aggregate tuning hours. Multiplying 18,859 times 12 times $0.0015, the SoundExchange fee is $339.46, but this is only the current SoundExchange cost. In Fall 2009 his stations were also paying $.37 per gigabyte of data transferred. One month in fall 2009, KNDE's billing cycle had 628 gigabytes (GB) of transfer. The 628 GB times $.37 came to a transfer cost of $232.36. In addition to the Sound-Exchange and data transfer variable costs, the station had about $200 in fixed costs. These three parts of Internet delivery of their stream to an audience in that one month came to $771.82 for an average of less than 100 listeners at a time. If a station assumes that most of the listening is from 6 a.m. to 10 p.m. 30 days a month, the average becomes 39 listeners at a time. This is where the economic challenge really comes in: if that listenership moves to 500 listeners, the calculation becomes (500*16*30) * 12 * $0.0015 for a SoundExchange cost of $4320. This calculation should take into account that the $0.0015 is the rate for 2009—the percentage is set to continue to increase. Adding the $4320 to bandwidth fees of $8051, and with 500 listeners, the station bears an expense of $12,571 for 480 hours of programming, at a cost of $26 dollars an hour.

There is an assumption that bandwidth rates, subject to free market forces, will continue to decrease—in 2009 they are 25% cheaper than they were in 2007. If bandwidth costs are cut in half by 2015, and with the 500 listeners example, the data transfer cost is down to $4025. Dusterhoff says that "however since (what is judged by the industry to be) a constitutionally questionable board sets the royalty rates, it will continue to increase" and would be $7200 based on the NAB settlement rate. All this should be evaluated within the context that local market radio stations continue to take actions that push listeners to listening via the Internet. Combine this trend with continued growth of Wi-Fi radios for the home and car, and the continued plans of automakers to increasingly make such radios standard or available as options in cars, more and more listening to traditional radio stations will be through the Internet instead of through traditional distribution, and the cost to reach listeners through the Internet delivery of the signal based on current costs are significantly higher than the costs to reach them through traditional over-the-air means.

The cost of the SoundExchange settlement could be absorbed by a local market radio station based on the old model of the Internet delivery of the signal being to a small number of listeners. However, when most of the audience listens via the Internet, the costs in even a relatively small market will reach half a million dollars a year, which stations cannot offset based on current advertising revenues. Specifically, Dusterhoff calculated that if the current listeners to KNDE FM in Bryan–College Station, Texas, currently listened on the Internet instead of over the air to the station, it would be costing the station, based on the current percentage, $465,000 a year. Dusterhoff believes radio stations will be forced to find a way to play fewer songs per hour to minimize their SoundExchange cost and that court decisions to provide relief will be necessary for local market broadcast radio stations to survive.

Streaming service provider Live365 has challenged the legality of the Copyright Royalty Board (CRB) and its ability to levy fees, particularly based on whether the judges of the CRB were appointed in violation of the separation of powers mandate of the U.S. Constitution. Live365's argument suggests the

CRB has "expansive executive authority . . . unsupervised by the Librarian of Congress or by any other Executive Branch official."[25] This lawsuit may take years to move through the court system.

Management Implications: Technology and Audience Synergies

Using an in-depth review of one local radio operation followed up by discussions with other radio managers shows this one case is a good representation of what is relevant about local market radio. What is obvious is that local radio is not as much about the technology as it is about the philosophy and implementation—a service to an audience. Interviews and comments from bloggers confirm an irritation from those who see radio as a commitment to serving their audience with what has happened to local market broadcasting because bottom-line mentality corporate ownership and efforts to run and/or buy and sell a local radio station based solely on measuring how to maximize profits without regard for serving an audience.[26] Similar comments are found in responses from radio managers summarized below. It seems the FCC got it right early on with the PICON principle (serve the public's interest, convenience and necessity), or at least ingrained a lot of people with these ideals of what service means.

Modern technology allows one person to handle more work and allows an efficiency in how programming is created and presented. The problem lies in how much work management can expect one person to do and the related need to be extremely efficient with the work employees do as a means of controlling personnel costs. Radio stations have naturally made adjustments in programming because of technology. It is easy now to interact with the audience through digital technologies. Listeners expect to be able to send text messages and offer opinions on a Web page, among other activities. Radio stations increasingly try to accommodate this audience interest by maintaining an interactive website and otherwise responding to the latest technological and social media trends. The problem with all this seems to be that, as Bryan Broadcasting General Manager Ben Downs says, they are "replacing old media dollars with new media pennies." Digital technologies allow an increasing number of services, that creates fewer listeners or users per service, and that translates into less revenue.

Results from a non-scientific survey of various radio industry managers, who often did not want to be identified because their views clash with corporate activities and strategies, suggest most people in the industry believe as bloggers noted above do that local radio can survive and thrive if operated properly. Mostly, that means focusing on serving the local community. There seems to be a belief that local radio stations engaged in a rededication to local programming (within what is similarly often called hyperlocal) can maintain significant audience interest in their stations in this period of increased digital competition from mobile media, satellite radio, Internet Radio and Wi-Fi radio. The belief is that there is still a place for local radio if it is truly local, with entertaining and compelling content. The problem is, according to these local market station managers, it is not really happening within the large group owners. One longtime major market radio manager and former Clear Channel employee said "no major broadcast group is currently engaged in a 'rededication to local programming.' To the contrary, the majority of corporate broadcast groups are utilizing distribution technology to disseminate centralized programming to their facilities. Local programming may consist of one or two dayparts of locally hosted content; the majority of the broadcast day is 'network' programming. That said, yes, local programming will always have an edge over mass produced programming. Unfortunately, while listeners will turn to local stations for information regarding severe weather or traffic or police emergencies, this does not convert the listener to a loyal partisan of the medium."

What strategies are local radio station managers using to maintain and/or build listenership in these days of increased digital competition? A commonly expressed sentiment was that they are doing very little to counteract the increased digital competition. One manager specifically said "local radio has become increasingly uninterested in serving the communities although they pay a lot of 'lip-service' to it."

Another noted that radio (in major markets especially) is having to reinvent itself daily due to the data gathered from its new rating system, the PPM, or Personal People Meter, and said the PPM allows stations to more quickly see patterns and trends in what the audience likes, and more quickly remove the elements they do not. But another radio manager says that while digital competition has siphoned listenership, ratings are still based on share of audience, and a high share while holding on to advertising rate integrity is still the goal. This manager says radio management is reacting to the new ratings methodology of PPM by cutting back on all on-air content and turning radio into jukeboxes, and that "the digital competition is a far more effective jukebox."

One of the assumptions in recent years is that local radio stations have seen significant declines in listenership in these days of increased digital competition. One manager said, "from what I have read and studied, we have not faced a real decline. While it is true that there are millions of people that subscribe to satellite radio, and millions more that download and play music via their computer, radio usage as a whole has not experienced the decline in audience as it is perceived in the media. That was a brilliant marketing campaign led primarily by satellite providers, who, by the way, were forced to merge themselves in order to stay afloat." Another noted, "how it happened was the clash of technology versus greed. Plain and simple. The combination of the Telecommunications Act that was 'supposed' to create more competition did the exact opposite. New technology allowed more stations to operate with less people, and while the programming was 'watered down' that was a conscious decision that was made, first and foremost to boost the bottom line, and alleviate the pressures to perform placed on radio station owners, managers and groups, who made outlandish financial and audience projections to make themselves, their stations and their companies more attractive to stockholders, and potential stockholders. Because of that, the creative process that made most radio stations magical, vibrant and exciting to listen to went away."

Others noted that digital competition is still low in penetration and generating share of audience, but is siphoning listenership in general. Radio station managers seem to agree that local stations should join the digital revolution and develop apps (applications) for distribution of broadcast programming online and via mobile phone. Growth of mobile broadband is seen as positive in that it can extend a local station's potential audience, and one manager noted that "more available listeners will never be a bad thing for a well programmed radio station."

Does HD radio have the potential to help local market radio stations? One manager said the programming is still in its infancy but that HD radio would evolve. It was noted that Europe has proven the technology is viable when done right, but once some technical and FCC issues in the United States are resolved, "HD radio will be more viable, profitable, and a great way for us to remain competitive in the ever changing landscape that is broadcasting." However, another manager said, "HD radio is an irrelevant distribution platform. It is limited in signal strength, and because the consortium agreed to make HD broadcasting commercial free, there has been no investment into the programming. Consequently, HD is used as a throw-away; much like FM was used in the early days. FM caught on because of the quality differential from AM and because there were no other distribution platforms available. HD is not significantly better than FM and there are more reliable options." It should be noted that in November 2009, iBiquity Digital Corporation and National Public Radio announced a deal that would allow an increase in power of HD Radio signals.[27]

In some of the final analysis of the current state of local market radio, one manager said radio usage is still huge, and it is still free, but that the only thing wrong with local radio is that in most cases it is not local: "personalities who are out of touch, and largely out of town will never make that local connection, and you can hear it." Another said that the big corporations are killing local radio and have been for years: "In fact, it is hardly local any-more—mostly syndication and satellite with some local voice-tracking. A lot of people still listen to 'local' radio but the money is never enough for the 'bean counters' running the stations these days and the future of local radio looks bleak because of it. There are very few people still doing local radio . . . the ones left in the business are over-worked and underpaid." Another's final analysis of the current state of local market radio was: "In a word, 'deregulation.' Once the 7/7 rule was waived and groups were allowed to consolidate stations, the industry began its downward spiral. The small owners could no longer compete with the mega companies. Groups of five or six stations in a single market would dominate the revenue and close the local operator with just one or two stations. Cost requirements quickly exceeded the need for creative excellence. In general, the talent in the industry has been replaced with less expensive, less experienced and less qualified entertainment."

Management Implications: Competing with New Technology

Internet radio has been around for a long time, having started in the mid-1990s as the Internet became established as a new media form. Running an Internet radio station is also known as webcasting, among other names, and refers to a constant stream of audio that users can listen to like a radio station. Internet radio stations may be information driven or have other content variations, but most Internet radio stations, as is the case with broadcast radio, have been music focused. Of course part of competing with the technology of Internet radio is for a local broadcast station to also stream its service over the Internet, and if the station has extra local HD channels (to be discussed more later), to also stream those channels. Early days of Internet radio were more problematic because of dial-up connections; high speed Internet has made listening to Internet radio easy.

Wi-Fi radio is an extension of Internet radio. While Internet radio is a description based upon how it is distributed, Wi-Fi radio refers to how it is received. With wireless broadband increasingly available and adopted in homes, Wi-Fi hot spots available in business and public locations nationwide, 3G cell phone services having Internet options, and the roll-out of city-wide wireless Internet called Wi-Max,[28] the growth of Wi-Fi radio has the potential to explode in growth in the coming years. This combination of broadband penetration, continued increases in capacity and speed, home and mobile Wi-Fi, and mobile media devices has the potential to make services like satellite and HD radio unnecessary.

What are the basics of satellite radio? The FCC opened up the S band (2.3 GHz) of the electromagnetic spectrum in the early 1990s for a nationwide Digital Audio Radio Service. Four companies participated in the bidding for these frequencies and the companies that ultimately became XM and Sirius each paid more than $80 million for the right to use the frequency.[29] Each launched their own set of satellites into geostationary orbit, built land-based translators to help with signal distribution, and established chipsets for receivers to receive their signals, with service starting in 2001 for XM and 2002 for Sirius. With both companies still losing money several years later, the Department of Justice and FCC approved a merger of the two companies in 2008; the merged company became Sirius XM. Customers still receive separate services since the two companies' equipment is not compatible, but customers of one service can buy a package that includes channels unique to the other service.

What are the basics of HD radio? HD radio is a proprietary technology owned by iBiquity. Local market stations pay a fee to acquire the HD transmission technology, then pay an annual fee as HD broadcast stations. HD refers to hybrid digital—a combination analog and digital signal is transmitted—and radio receivers that have HD reception circuitry are able to receive extra channels broadcast by the HD radio station. (Some publications also call it high definition, but iBiquity now says the HD stands for nothing—it is simply the name of their service.) HD radio is an in band on channel (IBOC) service, where the digital signal is added to the standard analog signal. HD stations on the FM band have an HD1 channel, that is a digital copy of the analog channel, then can have an HD2 channel as well as an HD3 channel. So, like HDTV, HD Radio on FM stations allows the local station to do a multicasting service. As an example from the Bryan Broadcasting stations, their analog and HD1 service on KNDE is a contemporary hits format called Candy 95, the HD2 channel is Rock Candy and the HD3 channel is a sports channel called "Play by Replay." iBiquity Digital Corporation's website notes that HD on AM Radio, even though multicasting is not possible, sounds like FM.[30] iBiquity also promotes iTunes tagging as a benefit of HD radio. To remember a song heard on participating HD radio stations, the user pushes the "Tag" button on an iTunes Tagging-enabled HD radio receiver. The receiver stores information regarding tagged selections for the next time the user's iPod is docked to the receiver.[31]

So how do these digital technologies relate to broadcast radio? A study conducted by Arbitron and Edison media in 2009, seventeenth in a series of studies conducted since 1998, measured how digital media is affecting terrestrial radio.[32] Some of the important basic facts regarding Internet radio as a threat to local market radio found in the most recent study include: (1) 85% of Americans have some Internet access from somewhere, (2) there has been significant transition from dial-up to broadband, where in 2005 half the population with Internet connections had dial-up and half had broadband, in 2009, 82% surveyed had broadband, (3) about half the population had listened to online radio and more than one-fourth had listened within a month prior to the survey and about 42 million people, or 17% of the 12+ population, had listened to online radio in the previous week, and (4) about one-third of those surveyed were interested in having Internet radio in the car. Other issues of digital media competition found in the study included: (1) the majority of the digital radio audience expects to listen to the same amount of local broadcast radio in the future, (2) even with more than 40% reporting owning an iPod or other MP3 player, only one in seven reported less radio time listening due to time spent with the device, (3) nearly two-thirds were aware of the merged Sirius XM Satellite Radio service, but most of those who were aware and do not have it had already decided they were not interested in the service, and (4) despite heavy promotion in recent years, only 29% reported having heard anything about HD Radio.

The final set of important information from this most recent study relates to the audience's attitude toward local broadcast radio. In a list of "platforms that have had a big impact on their lives," AM/FM radio was third on the list after cell phones and iPhones, well ahead of things like HD Radio, online radio, or audio podcasts, with results overall showing that cell phones first and AM/FM radio second (among the various devices and platforms included in the survey) have had the greatest impact of people's lives.

Future Implications

The future of local market radio is based on increased competition. A generation ago, a local radio station introduced the audience to new music and gave listeners a chance to record a rough copy of a favorite song. Now users can find the music they want via the Internet and download or make perfectly clean copies, often without cost. The dynamics have changed. But good local market radio is about more than music. It is about a service to an audience that includes news, sports, weather, emergency information, and a relationship. The Clear Channel style of streamlining operations gave the on-air sound a high quality but eliminated the relationship with the local listener. However, in this new world of competition, there will be fewer listeners to each service, and thus there probably needs to be some way to keep personnel costs down while maintaining local connections.

One way local stations have made programming more efficient is through syndicated shows by daypart, especially well served in the evening with shows like *Delilah, Lia* and *CMT Radio Live*. These programs run from 7 p.m. to 12 a.m. no matter which time zone the station is in. With *Delilah* as an example, the show is actually live for five hours (on the nights it is live—sometimes, especially weekends, there are replays that sound live) from 7:00 to midnight Central Time. The Eastern time zone gets the previous night's Central 11:00 to 12:00 hour of programming fed at 7:00-8:00 p.m. Eastern. The Mountain time zone gets the current night's 7:00-8:00 p.m. Central hour fed to them for their 11:00 to midnight hour, and the Pacific Time Zone gets that night's Central Times Zone 7:00 to 9:00 hours fed from ten to midnight Pacific. Technology allows this to be implemented seamlessly. Of course the greatest challenge lies in taking and airing live phone calls—the Central time zone always gets true live; the other time zones get illusions of live.

The original Satellite Music Network concept was more practical to meet today's needs than automation with voice-tracking, but with one exception. Syndicated shows as described above fit the daypart, no matter which time zone; 24/7 satellite-delivered formats cannot do this as easily. The problem with time zones can be resolved with one simple adjustment: with the technology as it is today, stations in one time zone should develop partnerships to syndicate one daypart to multiple stations all across the time zone. Just as syndicated and network shows easily integrate spots, jingles, liners, and local news, weather and traffic, the local stations would do so with this kind of partnership.

There is currently a "Country Legends" station in Houston, Texas. This format could be exported to any local market station across the Central time zone and made to sound local everywhere through strategic content insertions. Each local station could also choose to do its own local morning show, or afternoon drive show, and so forth, but with technical abilities as they are, to send the implemented format from one location to all the others. This kind of network arrangement solves the time zone challenges of syndicated shows and traditional satellite music networks, but still allows a local station to be local. Many of the networks offered by ABC Radio Networks in recent years actually had moved to voice tracking—so even a live radio service delivered via satellite for local market stations to air and insert local content into was not live. To do a good job of connecting with the audience, the model of morning show network TV news shows should be followed: combine truly live network feeds with local content on each station. This system takes advantage of today's technology while staying true to the PICON principle, if implemented correctly.

The local station would then use this cost efficiency to put their local staff to work "being local." The model we have come to know with local market TV is that the "faces" of the network-affiliated station are the news anchors.

They have taken a national network schedule of programs and made the station local by properly combining national programming with local content and connections. Local radio has been essentially doing the same thing, and if the overall station is programmed properly, could continue to find the efficiencies of a network-style operation while still being local.

One constant that comes out of comments from stations interviewed and bloggers commenting about radio is about keeping it local, and serving the public's interest, convenience, and necessity. It seems that tradition must be re-established and maintained for the local audience to care about local market radio. Smaller radio markets also seem to have a better chance of being able to do this, now, and in the future, just by the nature of how small communities have local community connection in a way large cities seem to not be able to mirror. Competition is part of a free market economy. In a time of new competition for local market radio stations, those that remain competitive with compelling content for an audience should be able to survive and thrive, unless future regulatory actions create unfair disadvantages.

Notes

1. Alan Albarran, Jennifer Horst, Tania Khalaf, John Phillip Lay, Michael Mc-Cracken, Bill Mott, Heather Way, Tonya Anderson, Ligia Garcia Bejar, Anna L. Bussart, Elizabeth Daggett, Sarah Gibson, Matt Gorman, Danny Greer, and Miao Guo, "What Happened to Our Audience? Radio and New Technology Uses and Gratifications Among Young Adult Users," *Journal of Radio Studies* 14, no. 2 (November 2007): 92–101.

2. Amy Jordan, Nicole Trentacoste, Van Henderson, Jennifer Manganello and Martin Fishbein, "Measuring the Time Teens Spend With Media: Challenges and Opportunities," *Media Psychology* 9, no. 1 (2007): 19–41.

3. Sonia Livingston, "Drawing Conclusions from New Media Research: Reflections and Puzzles Regarding Children's Experience of the Internet," *Information and Society* 22, no. 4 (2006): 219–230.

4. Veerle Van Rompaey, Keith Roe, and Karin Struys, "Children's Influence on Internet Access at Home: Adoption and Use in the Family Context," *Information, Communication & Society* 5, no. 2 (April 2002): 189–206.

5. Thorbjorn Broddason, "Youth and New Media in the New Millennium," *Nordicom Review* 27, no. 2 (2006): 105–118.

6. "US Broadband Penetration Grows to 63%," *The Bandwidth Report*, http://www.websiteoptimization.com/bw/ (27 July 2009).

7. Richard Berry, "Will the iPod Kill the Radio Star?" *Convergence: The International Journal of Research into New Media Technologies* 12, no. 2 (2006): 143–162.

8. Deborah Potter, "iPod, You Pod, We All Pod," *American Journalism Review* 28, no. 1 (February/March 2006): 64.

9. Anne Becker, "Multiplatform Summer," *Broadcasting and Cable* 136, no. 23 (June 5, 2006): 8. Also, Anthony Crupi and Mike Shields, "MTVN Offers Series on iTunes, Creates New Unit," *Mediaweek*, 16, no. 5 (30 January 2006): 4.

10. Michael Applebaum, "Test Ride," *Mediaweek*, 17, no. 14 (2 April 2007): 25–36.

11. Tom Lowry, "Antenna Adjustment," *BusinessWeek* 3938 (20 June 2005): 64-70. http://www.businessweek.com/magazine/content/05_25/b3938093_mz016.htm (10 August 2009). Also, Bill Moyers, "Politics and Economy: Virtual Radio" Transcript, http://www.pbs.org/now/transcript/transcript_clearc.html (18 June 2009).

12. Greater understanding about some of the early days of radio can be found on the *Documenting Early Radio* article at http://www.midcoast.com/~lizmcl/earlyradio .html and *United States Early Radio History* at http://earlyradiohistory.us/index.html.

13. Walter J. Beaupre, "Music Electrically Transcribed!" *OTR Articles*, http://www .otrsite.com/articles/artwb006.html (26 July 2009).

14. Carts were designed as tape loops that used special frequency tones recorded on the tape to designated starting points of program material among other functions.

15. Jim Price, "Cart Machines," *Pro Sound Resources*, http://www.jimprice.com/ prosound/carts.htm (20 June 2006).

16. See Music Master for Windows at http://www.mmwin.com/.

17. Bill Moyers, http://www.pbs.org/now/transcript/transcript_clearc.html (18 June 2009).

18. See *Tom Joyner Show* at http://www.blackamericaweb.com/?q=tjms.

19. Supplier Relations US, LLC, "Radio and Television Broadcasting and Wireless Communications Equipment Manufacturing Industry in the U.S. and its International Trade [Q3 2009 Edition]," http://www.marketresearch.com/map/prod/2432024.html (3 October 2009).

20. First Research, Inc., "Radio Broadcasting and Programming" (abstract), *Technology & Media* 2009, http://www.marketresearch.com (3 October 2009).

21. Steve McClellan, "The Karmazin Factor," *Broadcasting & Cable*, (17 November 1997), http://www.encyclopedia.com/doc/1G1-20011794.html (October 3, 2009).

22. Mark Ramsey, "Nielsen's Council for Research Excellence Blog," *Hear 2.0: Radio's Future* 2009, http://www.hear2.com/ (16 November 2009).

23. See the company's website at http://www.wireready.com/.

24. Olga Kharif, "The Last Days of Internet Radio?" *BusinessWeek Online*, http:// www.businessweek.com/technology/content/mar2007/tc20070307_534338.htm (28 September 2009).

25. "Live365 Files Lawsuit to Clarify Copyright Royalty Board Status," *Radiomag Online*, http://radiomagonline.com/IT_technology/streaming/libe365-sues-crb-20090901/ (28 September 2009).

26. See for example http://georgereedradiotv.blogspot.com/, http://blog.bia.com/ bia/, http://insidemusicmedia.blogspot.com/, http://ericrhoads.blogs.com/ink_tank/, and http://www.hear2.com/.

27. "NPR & iBiquity Strike Deal on HD Radio Power Increase," *iBiquity Digital*, http://www.ibiquity.com/press_room/news_releases/2009/1388 (16 November 2009).

28. See, for example, http://www.clear.com and http://www.wimax.com/.

29. See, for example, http://electronics.howstuffworks.com/satellite-radio1.htm; and http://www.carinsurance.com/Articles/content106.aspx.

30. See http://www.ibiquity.com/.

31. See http://www.ibiquity.com/press_room/fast_facts/all_about_itunes_tagging.

32. Tom Webster, "The Infinite Dial 2009," *Edison Research*, http://www.edison-research.com/home/archives/2009/04/the_infinite_dial_2009_presentation.php (28 September 2009).

Bibliography

Advanced Publishing, "Bosacks 5 Key Trends in Magazine Publishing—Clear Call to Action by Publishers," *Advanced Publishing*, 29 April 2008, blogs.advancedpublishing .com (26 September 2009).

Ahlers, Douglas, and John Hessen. "Traditional Media in the Digital Age: Data about News Habits and Advertiser Spending Lead to a Reassessment of Media's Prospects and Possibilities," *Nieman Reports*, (Fall 2005): 65–68.

Ahrens, Frank, "Gannett's Profit Plummets 53 Percent," *The Washington Post*, 20 October 2009.

Ala-Fossi, Marko, Piet Bakker, Hanna-Kaisa Ellonen, Lucy Küng, L., Stephen Lax, C. Sabada, and Richard van der Wurff, "The Impact of the Internet on Business Models in the Media Industries—A Sector by Sector Analysis," in *The Internet and the Mass Media*, ed. Lucy Kung, Robert G. Picard, and Ruth Towse. London: Sage, 2008, 149–169.

Albarran, Alan B. *Media Economics: Understanding Markets, Industries, and Concepts.* Ames, IA: Iowa State University Press, 2002.

Albarran, Alan B., Sylvia Chan-Olmsted, and Michael O. Wirth. *Handbook of Media Economics.* Mahwah, NJ: Lawrence Erlbaum Associates, 2006.

Alderman, John. *Sonic Boom: Napster, MP3, and the New Pioneers of Music.* New York: Basic Books, 2001.

Allen, Katie, "Heavy Tomes Turned into a Light Read with the Virtual Pocket Library: Publishers Expect Digital Ebooks to be the New Blockbusters," *The Guardian*, 10 September 2009.

Altschuler, Glenn C. *All Shook Up: How Rock 'n' Roll Changed America* (New York: Oxford University Press, USA, 2003).

Anderson, Chris. *The Long Tail: Why the Future of Business is Selling Less of More.* New York: Hyperion, 2006.

Anderson, Chris. *Free: The Future of a Radical Price.* New York: Hyperion, 2009.

Arbitron, "The Infinite Dial 2009," http://www.arbitron.com/study/digital_radio_study.asp (15 October 2009).

Atran, Scott. "The Trouble with Memes: Inference Versus Imitation in Cultural Creation." *Human Nature* 12, no. 4 (2006): 351–381.

Bass, Frank. "A New Growth Model Product for Consumer Durables." *Management Science* 15, no. 5 (2969): 215–227.

Bates, Benjamin J., "Transforming Information Markets: Implications of the Digital Network Economy," *Proceedings of the American Society for Information Science and Technology* 45: 2009, 11.

Bellamy, Robert V., Jr., and James R. Walker. *Television and the Remote Control: Grazing on a Vast Wasteland.* New York: Guilford, 1996.

Berry, Richard. "Will the iPod Kill the Radio Star?" *Convergence: The International Journal of Research into New Media Technologies* 12, no. 2 (2006): 143–162.

Bivings Group—"American Newspapers and the Internet: Threat or Opportunity," http://www.bivingsreport.com/2007/american-newspapers-and-the-internet-threat-or-opportunity/ (10 January 2010).

Blankenship, Donna Gordon, "Seattle Post-Intelligencer Newspaper Goes Web-Only," *Yahoo! Tech News,* 17 March 2009, http://tech.yahoo.com/news/ap/20090317/ap_on_hi_te/seattle_p_i (24 March 2009).

Bowman, Shayne, and Chris Willis. "We Media: How Audiences are Shaping the Future of News and Information," http://www.hypergene.net/wemedia/download/we_media.pdf (10 January 2010).

Bradley, Tony."Kindle PDF Support Broadens Ebook Reader Appeal for Businesses," *BizFeed,* 24 November 2009, www.pcworld.com (5 December 2009).

Bradshaw, Paul, "Making Money from Journalism: New Media Business Models (A Model for the 21st Century Newsroom pt5), *Online Journalism Blog* 2008, http://onlinejournalismblog.com/2008/01/28/making-money-from-journalism-new-media-business-models-a-model-for-the-21st-century-newsroom-pt5/ (20 Sept. 2009).

Broddason, Thorbjorn, "Youth and New Media in the New Millennium." *Nordicom Review* 27, no. 2 (2006): 105–118.

Brown, Paul, "Ebooks Battle for Next Chapter: Publishers Are Now Willing to Embrace Ebooks—but Are They Ready to Head Off the Threat of a Format War?" *The Guardian,* 23 April 2009, Technology Section, 6.

Burt, Ronald S. *Contagion: Models of Imitation and Interpersonal Influence.* (2004): http://faculty.chicagobooth.edu/ronald.burt/teaching/Contagion.pdf (15 November 2009).

Carl, Walter J., "The Role of Disclosure in Organized Word-of-Mouth Marketing Programs." *Journal of Marketing Communications* 14, no. 3 (2008): 225–241.

Carroll, Sean B., "Evolution at Two Levels: On Genes and Form." *PLoS Biology* 3, no. 7 (2005): e245, http://www.ncbi.nlm.nih.gov/pmc/articles/PMC1174822/?tool=pmcentrez (14 November 2009).

Carton, Sean, "Five Rules of Viral Marketing." *ClickZ,* February 5, 2007, http://www.clickz.com/3624847/ (8 December 2009).

Caves, Richard E. *Switching Channels*. Cambridge, MA: Harvard University Press, 2005.

Chang, Jeff, and D.J. Kool Herc. *Can't Stop Won't Stop: A History of the Hip-Hop Generation*. New York: Picador, 2005.

Comiskey, Barrett, J.D. Albert, Hidekazu Yoshizawa, and Joseph Jacobson, "An Electrophoretic Ink for All-Printed Reflective Electronic Displays." *Nature*, 394 (16 July 1998), 253–255.

Dawkins, Richard. *The Selfish Gene*. New York: Oxford University Press, 1976.

"Definition Of: Connectivity." *PC Mag.com*. http://www.pcmag.com/encyclopedia_term/0,2542,t=connectivity&i=40241,00.asp/ (16 November 2009).

Dimmick, John. *Media Competition and Coexistence: The Theory of the Niche*. Mahwah, NJ: Lawrence Erlbaum Press, 2003.

Dobele, Angela, David Toleman, and Michael Beverland, "Controlled Infection! Spreading the Brand Message through Viral Marketing." *Business Horizons* 48, no. 2 (March-April 2005): 143–149.

Donath, J., and D. Boyd. "Public Displays of Connection." *BT Technology Journal* 22, no. 4 (2004): 71–82.

Dong-Seok, Sah, "Saving Newspapers," *Korea Times*, 8 April 2009.

Duderstadt, James J., "Will Google's Book Scan Project Transform Academia?" (online video), 2009, www.youtube.com (12 December 2009).

Edmonds, Bruce, "The Revealed Poverty of the Gene-Meme Analogy—Why Memetics per se has Failed to Produce Substantive Results." *Journal of Memetics—Evolutionary Models of Information Transmission* 9, no. 1 (2002), http://cfpm.org/jom-emit/2005/vol9/edmonds_b.html (18 November 2009).

Eisenstein, Elizabeth L. *The Printing Revolution in Early Modern Europe*. Cambridge, England: Cambridge University Press, 1983.

Eldred, Eric, et al., Petitioners, v. John D. Ashcroft, Attorney General, 537 U.S. 186.

Ellison, N. B., C. Steinfeld, and C. Lampe, "The Benefits of Facebook 'Friends': Social Capital and College Students' Use of Online Social Network Sites." *Journal of Computer-Mediated Communication* 12, no. 4 (2004).

Evans, Jonny, "Apple Tablet Will Revolutionize Ebook Publishing," *9 to 5 Mac: Apple Intelligence*, 30 September 2009. www.9to5mac.com (18 December 2009).

Fallows, James. *Breaking the News: How the Media Undermine American Democracy*. New York: Pantheon Books, 1996.

"Fat Newspaper Profits Are History," *Reflections of a Newsosaur*, 21 October 2008. newsosaur.blogspot.com (15 December 2009).

Friedman, Wayne, "Double Play: Big Web Users are Big TV Viewers." *Media Daily News*, October 31, 2008, http://www.mediapost.com/publications/?fa=Articles.showArticle&art_aid=93900 (10 January 2010).

Ferguson, Douglas A., "Industry Specific Management Issues," in *Handbook of Media Management and Economics* ed. Alan Albarran, Sylvia Chan-Olmsted, and Michael Wirth, 297–323. Mahwah, NJ: Lawrence Erlbaum, 2005.

Feenstra, Johan, and Rob Hayes. *Electrowetting Displays*. Eindhoven, the Netherlands: Liquavista, 2006.

Forrester, Chris. *The Business of Digital Television*. Oxford: Focal Press, 2000.

Füssel, Stephan. *Gutenberg and the Impact of Printing*. Aldershot, Hampshire, England: Ashgate Publishing, 2005.

Gallagher, Victoria, "Bookseller Survey: Cheaper E-books Needed to Drive Digital Growth," *The Bookseller*, 12 February 2009. www.thebookseller.com (9 December 2009).

Garfield, Bob. *The Chaos Scenario*. Nashville: Stielstra Publishing, 2009.

Gers, Matt, "The Case for Memes." *Biological Theory*, 3, 4 (Fall 2008): 305–315.

Gill, Kathy, "Publishing Industry Responds to Digital Disruption by Delaying Ebooks," *MCDM: Flip the Media*, 9 December 2009. flipthemedia.com (9 December 2009).

Ginsberg, Jane C., "Copyright and Control Over the New Technologies of Dissemination," *Columbia Law Review*, Vol. 101, no. 7 (Nov. 2001): 1613–1647.

Granovetter, Mark S., "The Strength of Weak Ties." *The American Journal of Sociology* 78, no. 6 (May 1973): 1360–1380.

Grimes, Robert G. "General Semantics and Memetics: A Tentative Relationship?," *Et Cetera* 55, no. 1 (1998): 30–33.

Graham-Rowe, Duncan, "Paper Comes Alive," *New Scientist* 179, no. 2414, 16–17.

Green, James N., "From Printer to Publisher: Mathew Carey and the Origins of Nineteenth-Century Book Publishing." Pp. 26–44 in *Getting the Books Out: Papers of the Chicago Conference on the Book in 19th-Century America*, edited by Michael Hackenberg. Washington: The Center for the Book, Library of Congress, 1987.

FCC, "FCC's Review of Broadcast Station Ownership Rules," (2009), http://www.fcc.gov/cgb/consumerfacts/reviewrules.html (1 Nov. 2009).

Fleishman, Glenn. "Dual Perspectives, Future of Newspapers: Newspapers Founder, But Civic Journalism May Survive." *Wired*, 14 July 2009, http://www.wired.com/dualperspectives/article/news/2009/07/dp_newspaper_ars0714 (November 2009).

Grant, August E., and Jennifer H. Meadows (eds.). *Communication Technology Update and Fundamentals*, 11th ed. Burlington, MA: Focal Press, 2008.

Harries, Dan, ed. *The New Media Book*. London: British Film Institute, 2002.

Hartley, John. *Creative Industries*. Oxford, UK: Blackwell Publishing, 2005.

Heylighen, Francis, "Objective, Subjective and Intersubjective Selectors of Knowledge." *Evolution and Cognition* 3, no. 1 (1997): 63–67.

Hoechsmann, Michael and Giuliana Cucinella, "My Name is Sacha: Fiction and Fact in a New Media Era." *Taboo* (Spring-Summer 2007): 91–97.

Holton, Lisa, "Trendspotting 2009: Lisa Holton's Predictions," *Publishing Trends*, January 2009. www.publishingtrends.com (26 September 2009).

Howard, Nicole. *The Book: The Life Story of a Technology*. Westport, CT: Greenwood Press, 2005.

"Industry Statistics," *International Digital Publishing Forum*, n.d. http://www.openebook.org/ (9 December 2009).

Jacobson, Joseph M., Barrett Comiskey, C. Turner, J. Albert, and P. Tsao, "The Last Book," *IBM Systems Journal* 36, no. 3 (1997), 457–463.

Jenkins, Henry. *Convergence Culture, Where Old and New Media Collide*. New York: New York University Press, 2006.

Jones, Jonathan, "What You Should Know About the Newspaper Industry," *PBS Frontline Newswar*, 27 February 2007, www.pbs.org/wgbh/pages/frontline/newswar (15 December 2009).

Jones, Sydney, and Susannah Fox. *Generations Online in 2009*. Pew Internet & American Life Project. http://www.pewinternet.org/~/media//Files/Reports/2009/PIP_Generations_2009.pdf (4 December 2009).

Jordan, Amy, Nicole Trentacoste, Vani Henderson, Jennifer Manganello, and Martin Fishbein, "Measuring the Time Teens Spend with Media: Challenges and Opportunities." *Media Psychology* 9, no. 1 (2007): 19–41.

Joyce, Elizabeth, and Robert E. Kraut, "Predicting Continued Participation in Newsgroups." *Journal of Computer Mediated Communication* 11, no. 3 (2006): 723–747.

Kahney, Leander. *The Cult of iPod*. San Francisco: No Starch Press, 2005.

Karakas, Fahri. "Welcome to World 2.0: The New Digital Ecosystem." *Journal of Business Strategy* 30, no. 4 (2009): 23–30.

Keith, Michael C. *The Radio Station: Broadcast, Satellite and Internet*, 8th ed. Burlington, MA: Focal Press, 2010.

Kelly, Joseph F. *The World of the Early Christians*. Collegeville, MN: The Order of St. Benedict, Inc., 1997.

Kinsley, Michael, "Life After Newspapers," *The Washington Post*, 6 April 2009, A15.

Kolb, Erik B., and Tuna N. Amobi. "Broadcasting, Cable & Satellite." Pp. 1–33 in *Standard & Poor's Industry Surveys*. New York: Standard and Poor's Equity Research Services, 2009.

Knopper, Steve. *Appetite for Self-Destruction: The Spectacular Crash of the Record Industry in the Digital Age*. New York: Free Press, 2009.

Küng, Lucy, Robert G. Picard, and Ruth Towse. *The Internet and The Mass Media*. London: Sage, 2008.

Lasica, J. D. *Online Journalism Review*, August 2003.

Lean, Geoffrey, "Mobile Phone Use Raises Children's Risk of Brain Cancer Fivefold," *The Independent*, September 21, 2008, http://www.independent.co.uk/news/science/mobile-phone-use-raises-childrens-risk-of-brain-cancer-fivefold-937005.html (15 January 2010).

Lee, Paul S. and Louis Leung, "Assessing the Displacement Effects of the Internet." *Telematics & Informatics* 25, no. 3 (2008). 145–155.

Lieberman, David, "Tension Mounts in E-Reader Saga: Publishers Aren't Happy with Amazon's Pricing," *USA Today*, December 11, 2009, 1B–2B.

Lister, Martin, Jon Dovey, Seth Giddings, Iain Grant and Kieran Kelly. *New Media: A Critical Introduction*. London: Routledge, 2003.

Manjoo, Farhad, "Fear the Kindle: Amazon's Amazing E-book Reader is Bad News for the Publishing Industry," *Slate*, 26 February 2009. www.slate.com (9 December 2009).

Manovich, Lev. *The Language of New Media*. Cambridge, MA: The MIT Press, 2001.

McKenna, Katelyn Y. A., Amie S. Green, and Marci E. J. Gleason, "Relationship Formation on the Internet: What is the Big Attraction?" *Journal of Social Issues* 58, no. 1 (2002): 9–31.

McMillan, Sally J., and Margaret Morrison, "Coming of Age with the Internet: A Qualitative Exploration of How the Internet Has Become an Integral Part of Young People's Lives." *New Media & Society* 8, no. 1 (2006): 73–95.

Miller, Mark K., "A Brief History of Broadcasting and Cable," *Broadcasting and Cable Yearbook 2009.* New Providence, NJ: ProQuest LLC, 2009.

Miller, Mark K., "A Chronology of the Electronic Media." *Broadcasting and Cable Yearbook 2009.* New Providence, NJ: ProQuest LLC, 2009.

Musser, Charles. *The Emergence of Cinema: The American Screen to 1907 (History of the American Cinema, Vol. 1),* (Berkeley, CA: University of California Press, 1994),

Negroponte, Nicholas. *Being Digital.* New York: Vintage Books, 1995.

Newlyn, Korbin, "History of Cell Phones—How This Omnipresent Device Progressed to Where It Is Today," http://EzineArticles.com/?expert=KorbinNewlyn (18 January 2010).

Nguyen, An, and Mark Western. "The Complementary Relationship Between the Internet and Traditional Mass Media: The Case of Online News and Information." *Information Research* 11, no. 3 (April 2006), http://informationr.net/ir/11-3/paper259.html (10 January 2010).

O'Reilly, Tim, "Why Kindle Should Be an Open Book," *Forbes,* 23 February 2009. www.forbes.com (16 December 2009).

Paczkowski, John, "Apple Pitching Tablet to Publishing Industry; Spring Launch Expected," *Digital Daily,* 9 Dec 2009, digitaldaily.allthingsd.com (9 December 2009).

Padgett, Tim. "What's Next After that Odd Chicken?" *Time,* October 8, 2004, 164.

Palmer, Shelly. *Television Disrupted: The Transition from Network to Networked TV.* (Burlington, MA: Focal Press, 2006).

Pastore, Michael, "Will Amazon.com Conquer the Ebook Industry?" *Epublishers Weekly,* 5 May 2009, epublishersweekly.blogspot.com (12 December 2009).

Pavlik, John V. *Media in The Digital Age.* New York: Columbia University Press, 2008.

Peacock, Marisa, "Ebook Publishing Trends: 70% Ready to Go Digital," *CMS Wire,* 15 October 2008, www.cmswire.com/ (9 December 2009).

Penenberg, Adam L. *Viral Loop.* New York: Hyperion, 2009.

Peters, Benjamin, "And Lead Us Not into Thinking the New is New: A Bibliographic Case for New Media History." *New Media & Society,* (2009), www.columbia.edu/~bjp2108/blog/Peters%20NMS%202009.pdf (12 January 2010).

Porter, Lance, and Guy J. Golan. "From Subservient Chickens to Brawny Men: Comparison of Viral Advertising to Television Advertising." *Journal of Interactive Advertising* 6, no. 2 (Spring): 30–38.

Potter, Deborah. "iPod, You Pod, We All Pod," *American Journalism Review* 28, no. 1, (February/March 2006): 64.

Putnam, Robert D. *Bowling Alone: The Collapse and Revival of American Community.* New York: Simon & Schuster, Inc, 2000.

Quick, Chris, "With Smartphone Adoption on the Rise, Opportunity for Marketers is Calling," *Nielsenwire,* 2009, http://blog.nielsen.com/nielsenwire/online_mobile/with-smartphone-adoption-on-the-rise-opportunity-for-marketers-is-calling/ (22 October 2009).

"Read it and Weep: Why Ebooks Must Change the Record," *The Business*, 26 January 2008.

Reardon, Kathleen K., and Everett M. Rogers, "Interpersonal Versus Mass Communication: A False Dichotomy." *Human Communication Research* 15, no. 2 (1988): 284–303.

Redden, Elizabeth, "Toward an All E-Textbook Campus," *Inside Higher Ed*, 2009. www.insidehighered.com (14 January 2009).

Rodowick, David Norman. *The Virtual Life of Film* (Cambridge, MA: Harvard University Press, 2007).

Rogers, Everett M. *Diffusion of Innovations*. New York: Free Press, 1962.

Rompaey, Veerle Van, Keith Roe, and Karin Struys, "Children's Influence on Internet Access at Home: Adoption and Use in the Family Context," *Information, Communication & Society* 5, no. 2 (April 2002): 189–206.

Rosen, Jay, "Emerging Alternatives," *Columbia Journalism Review (CJR)*, September/October 2003. (http://cjrarchives.org/issues/2003/5/alt-rosen.asp).

Rothman, David, "Trendspotting 2009: David Rothman's Predictions," *Publishing Trends*, January 2009. www.publishingtrends.com (26 September 2009).

Rushkoff, Douglas. *Media Virus!: Hidden Agendas in Popular Cultures*. New York: Ballantine Books, 1994.

Russo, Julie L., "User-Penetrated Content: Fan Video in the Age of Convergence." *Cinema Journal* 48, no. 4 (Summer 2009): 119–131.

Samuelson, Pamela, "The Generativity of *Sony v. Universal*: The Intellectual Property Legacy of Justice Stevens." *Fordham Law Review* 74, 1831 (2006): 101–145.

Schnittman, Evan, "Why Ebooks Must Fail," *Black Plastic Glasses: Musings on Publishing and Life in the Digital Age*, 30 March 2009, www.blackplasticglasses.com (9 December 2009).

Shao, Guosong, "Understanding the Appeal of User-Generated Media from the Users and Gratifications Perspective." *Internet Research* 19, no. 1 (2009): 7–25.

Shaver, Dan, and Mary Alice Shaver. "Directions for Media Management Research in the 21st Century." Pp. 639–54 in *Handbook of Media Management and Economics*, edited by Alan B. Albarran, Sylvia M. Chan-Olmsted, and Michael O. Wirth. Lawrence Erlbaum Associates, 2006.

Shuen, Amy. *Web 2.0: A Strategy Guide*. Sebastopol, CA: O'Reilly Media, 2008.

Shy, Oz. *The Economics of Network Industries*. New York: Cambridge University Press, 2001.

Smith, Sarah M., and Dean M. Krugman, "Viewer as Media Decision-Maker: Digital Video Recorders and Household Media Consumption," *International Journal of Advertising* 28, no. 2 (2009), 231–55.

Sterling, Christopher H., and John Michael Kittross. *Stay Tuned: A History of American Broadcasting*, 3rd. ed. Mahwah, NJ: Erlbaum, 2002.

Sylvie, George, and Patricia D. Witherspoon. *Time, Change and the American Newspaper*. Lawrence Erlbaum and Associates, Mahwah NJ, 2001.

Trammell, K., and A. Keshelashvili, "Examining New Influencers: A Self-Presentation Study of A-list Blogs." *Journalism and Mass Communication Quarterly* 82, no. 4 (2005): 968–982.

Trenholm, Rich, "Top Publisher Predicts 'Ebooks Will Sink the Publishing Industry,'" *Cnet UK*, 7 April 2009, crave.cnet.co.uk (9 December 2009).

TV Bureau of Advertising, Inc. "TV Basics: Television Households." http://www.tvb .org/rcentral/mediatrendstrack/tvbasics/02_TVhouseholds.asp (1 Nov. 2009).

"U.S. Bill Seeks to Rescue Faltering Newspapers," *Reuters*, 24 March 2009.

Viral Video Marketing Survey: The Agency Perspective. Los Angeles, CA: Feed Company. Feed Company. http://www.feedcompany.com/wp-content/uploads/Feed_ Company_Viral_Video_Marketing_Survey.pdf (2 December 2009).

Walker, James R., and Robert V. Bellamy, Jr., eds. *The Remote Control in the New Age of Television.* Westport, CT: Praeger, 1993.

Wasser, Frederick. *Vini, Vidi, Video: The Hollywood Empire and the VCR.* Austin, TX: University of Texas Press, 2002.

Waterman, David, "The Economics of Media Programming." Pp. 387-416 in *Handbook of Media Management and Economics.* Edited by Alan B. Albarran, Sylvia M. Chan-Olmsted, and Michael O. Wirth. Lawrence Erlbaum Associates, 2006.

Webster, James, "Audience Behavior in the New Media Environment," *Journal of Communication*, 36(1986): 77–91.

Weingarten, Matt, "Writer Speaks to Industry's Decline," *Cosmos Magazine*, 2009. www.cosmosmagazine.com (19 February 2009).

Weprin, Alex. "Much A-Twitter about Something." *Broadcasting & Cable*, 11 May 2009, 12–13.

Winograd, Morley, and Michael D. Hais. *Millennial Makeover: MySpace, YouTube, and the Future of American Politics* New Brunswick, NJ: Rutgers University Press, 2008.

Wolk, Douglas, "Dual Perspectives, Future of Newspapers: Profitless? Go Wireless." *Wired*, 14 July 2009, http://www.wired.com/dualperspectives/article/news/2009/07/ dp_newspaper_wired0714 (10 January 2010).

Woudstra, Wendy J., "The Future of Textbooks: Ebooks in the Classroom," *Publishing Central*, n.d. publishingcentral.com (7 December 2009).

Young, Jeffrey R., "How a Student-Friendly Kindle Could Change the Textbook Market," *The Chronicle of Higher Education*, 6 May 2009.

Zimmerman, Martin, "Story Not All Bleak for Newspaper Industry's Outlook," *Los Angeles Times*, 31 August 2009.

Online Resources

Berkman Center for Internet & Society, Harvard Law School, http://cyber.law .harvard.edu/research/internetdemocracy#

Journalism That Matters—http://journalismthatmatters.wordpress.com/resources/

Kennesaw Summit: Building a Charter for a Civic Journalism Professional Society http://kennesawsummit.kennesaw.edu/index.htm

Nielsen/Netratings—http://www.nielsen-netratings.com/

The Next Newsroom Project—http://www.nextnewsroom.com

Pew Center for Civic Journalism: http://www.pewcenter.org/doingcj/speeches/s_
spjheadline.html
Pew Internet & American Life Project—www.pewinternet.org
Poynter Institute—http://www.poynter.org
Pressthink—http://journalism.nyu.edu/pubzone/weblogs/pressthink/
Principles of Citizen Journalism Project: http://www.kcnn.org/principles
Project for Excellence in Journalism / Pew Research Center—State of the News Media
2008—http://www.stateofthenewsmedia.org/2008/
Society for New Communications Research—http://sncr.org

Index

Note: Page numbers followed by *f* indicate figures and by *t* indicate tables.

About the Editor

John Allen Hendricks (Ph.D., University of Southern Mississippi) is Director of the Division of Communication and Contemporary Culture and Professor of Communication at Stephen F. Austin State University in Nacogdoches, Texas. Dr. Hendricks worked in the radio industry for nineteen years. He worked in commercial radio for seven years and then served for twelve years as faculty director of a non-commercial, university-owned radio station providing oversight for its daily operation including programming, budgeting, and personnel.

Dr. Hendricks co-edited the books *Communicator-in-Chief: How Barack Obama Used New Media Technology to Win the White House* (with Robert E. Denton, Jr.) and *Techno Politics in Presidential Campaigning: New Voices, New Technologies, and New Voters* (with Lynda Lee Kaid). His research has also been published as scholarly articles and chapters in *21st Century Communication: A Reference Handbook; The Business of Entertainment: Movies, Music, and Television; American Journalism: History, Principles, Practices; Feedback; Journal of Radio and Audio Media; Southwestern Mass Communication Journal; Communications and the Law*, and numerous other publication outlets. Dr. Hendricks is editor of the Studies in New Media series for Lexington Books in Lanham, Maryland.

Dr. Hendricks currently serves as a member of the Board of Directors of the Broadcast Education Association (BEA) representing District 5. He is past President of the Oklahoma Broadcast Education Association (OBEA) and is a former ex officio member of the Board of Directors of the Oklahoma

Association of Broadcasters (OAB). He is past Chair of the Southern States Communication Association's (SSCA) Political Communication and Mass Communication divisions. Dr. Hendricks serves on the Editorial Advisory Board of the *Journal of Radio and Audio Media* and is a past member of the Editorial Advisory Board of the *Southwestern Mass Communication Journal.*

About the Contributors

Alan B. Albarran (Ph.D., The Ohio State University) is Director of the Center for Spanish Language Media and a Professor in the Department of Radio, Television and Film at the University of North Texas, in Denton, Texas. Dr. Albarran previously served as Editor of *The International Journal on Media Management* (2006-2008) and Editor of *The Journal of Media Economics* (1997-2005).

Dr. Albarran has authored/edited ten books, the most recent including *The Media Economy* (in press), *The Handbook of Spanish Language Media* (2009), *The Media as a Driver of the Information Society* (2009), *Management of Electronic Media*, 4th edition (2009), and *The Handbook of Media Management and Economics* (2006). In addition, he has published numerous articles in scholarly journals and chapters in edited volumes. He is finishing a new book entitled *The Transformation of the Media and Communication Industries*.

Internationally recognized as one of the leading scholars in the field of media management and economics, Dr. Albarran has lectured and presented workshops in a total of 20 countries throughout Europe, Asia, and Latin America. He also serves as an industry consultant and is designated as a leader with the Gerson Lehrman Council of Advisors. Dr. Albarran's recent awards include recognition as the Broadcast Education Association's Distinguished Scholar Award (2009), the *Journal of Media Economics* Award of Honor (2008), a Fulbright Senior Scholar Award (2006), and the Toulouse Scholar Award recognizing him as the outstanding member of the graduate faculty from the University of North Texas (2006). He previously served as the President of the Broadcast Education Association and the Texas Association of Broadcast Educators.

Robert Bellamy (Ph.D., University of Iowa) is an Associate Professor in the Department of Journalism and Multimedia Arts at Duquesne University. He has published widely in the areas of media and sports, impact of new technology on media industries, and television programming. Dr. Bellamy's most recent book is *Center Field Shot: A History of Baseball on Television* (University of Nebraska Press, 2008), co-authored with James R. Walker.

Alexander Cohen (Ph.D., State University of New York at Buffalo) joined the faculty at UC Berkeley in 1990, and has taught courses on the philosophy and rhetoric of science, film and the Internet, and the critical theory of technology. Dr. Cohen is a senior inventor for Intellectual Ventures and managing director at IDEON Partners, where he focuses on software patent analysis, and portfolio analysis and management for clients. In the past he was chairman and co-founder of Undergroundfilm.org, an online, nonprofit website dedicated to connecting independent filmmakers with their audiences via digital distribution technology, worked for Dreamworks and Imagine Entertainment on Pop.com, an early digital video website, and has worked for a variety of digital media–related Silicon Valley startups including Netscape, Excite@Home, and CNET among others.

Tony R. DeMars (Ph.D., University of Southern Mississippi) is Associate Professor and Radio-Television Division Director at Texas A&M University–Commerce. His teaching areas include broadcast journalism, new media technologies and broadcast production. His publications include the book *Modeling Behavior From Images of Reality in Television Narratives*, the *Journal of Radio and Audio Media* article "Buying Time to Start Spanish-Language Radio in San Antonio: Manuel Davila and the Development of Tejano Programming," and the book chapter "News Convergence Arrangements in Smaller Media Markets," in *Understanding Media Convergence: The State of the Field.*

Dr. DeMars is Council of Divisions Vice Chair in the Association for Education in Journalism and Mass Communication (AEJMC), Chair of the Production Aesthetics and Criticism Division in the Broadcast Education Association (BEA), Student Audio Chair for the BEA Festival, and is past president of the Texas Association of Broadcast Educators (TABE).

Douglas A. Ferguson (Ph.D., Bowling Green State University) is a Professor at the College of Charleston. He graduated from The Ohio State University with both bachelor's and master's degrees in Communication. He spent 13 years as a programmer and manager of an NBC television affiliate in Ohio before getting his doctorate degree in mass communication. After a dozen years teaching in Ohio, Dr. Ferguson went to South Carolina as the inaugural chair of the Department of Communication at the College of Charleston, later serving as the department's first graduate director. He is a former editor of the *Journal of Radio and Audio Media*. He has co-written three books on the television industry and has contributed seventeen research articles to refereed journals. Dr. Ferguson is interested in how newer media technologies influence traditional media.

Robert Gross (B.G.S., University of Michigan) is a graduate assistant in the Department of Journalism and Multimedia Arts at Duquesne University while pursuing his master's degree. His primary research interest is the recorded music business, which is particularly germane to his other profession as a songwriter/musician.

Jen McClure (M.A., Stanford University) is a renowned expert in new media and communications. Her career spans 25 years and includes experience in all facets of professional communications, including journalism, market and media research, media relations, public relations, strategic communications, print and online publishing and broadcast media. She is currently the chief marketing officer and director of community development for Redwood Collaborative Media and is the founder and president of the Society for New Communications Research (SNCR), a global nonprofit research and education foundation and think tank that focuses on the latest developments in media and communications.

Jennifer Meadows (Ph.D., University of Texas) is Professor and Head of the Media Arts Option in the Department of Communication Design at California State University, Chico. Her doctorate is in Radio, Television, Film and her master's degree is in Radio, Television, and Motion Pictures from the University of North Carolina, Chapel Hill. Dr. Meadows' research interests include social uses of technology, new media and entertainment and the changing audience and media industry behavior in the new media marketplace. She is the co-editor of the *Communication Technology Update & Fundamentals*, now in its 12th edition.

Steven Phipps (Ph.D., University of Missouri at Columbia) is a media researcher and historian currently teaching in the Communication program at Maryville University in St. Louis, Missouri. He has also taught at the

University of Missouri at St. Louis, Washington University in St. Louis, and Indiana University–Purdue University at Ft. Wayne. Dr. Phipps' research interests include media technology and revisionist media history. In his research he attempts to identify current trends and use them as a basis for predicting the future of the media. The subjects of his specific media research projects are varied, and have included such topics as early film history, the formulation and future of broadcast regulation, the role of women in WWII radio journalism, and, currently, the relationship between social media and media literacy.

Another strong interest, which is reflected in much of his teaching, is the current switch from analog to digital technology in many areas of the media. He is especially concerned with the role played by technological determinism in shaping human communication options and processes in the future, especially through those media that were formerly known as print and broadcast.

His published research has included articles in various academic journals in the field, including the *Journal of Broadcasting and Electronic Media*, *Journalism Quarterly*, and *Historical Journal of Film, Radio and Television*. In addition, his media research projects have resulted in conference papers, encyclopedia articles, and an appearance on the PBS television series *History Detectives* as a media researcher. Much of his published research concerns media technological history, media legislation and regulation, and the early history of radio.

Mary Jackson Pitts (Ph.D., University of Southern Mississippi) is a Professor in the College of Communications at Arkansas State University (ASU) and is the former editor of the *Southwestern Mass Communication Journal*. She worked in radio and television in the state of Arkansas prior to beginning her teaching career at ASU. Dr. Pitts is a media effects researcher who has focused much of her research on how traditional media are using digital media to further their scope and purpose. She has publications that address the changes the Internet has brought to the world of storytelling in television, and examined how traditional radio has used the Internet to promote the over the air product. Her most recent publication focuses on the use of podcasting and vodcasting and was published in the BEA *Feedback* publication this spring. She recently presented research at the Western Social Science Association (WSSA) that examined radio stations' use of the Internet as a public relations tool. Dr. Pitts' research also focuses on the impact new media has had on copyright issues.

Dr. Pitts serves as the faculty advisor for the College of Communications undergraduate student research association. She currently serves as the Vice

Chair of the Public Relations Division of the Southern States Communication Association (SSCA) and serves as co-chair of the Mass Communication Division of the Western Social Science Association (WSSA).

Susan Smith (M.Comm., Georgia State University) is an Assistant Professor of Telecommunications at Ball State University. Smith joined the staff after more than 24 years of experience in the news/communications industry. Her most recent work experience came at CNN.com, where she worked as a senior producer with the website's live video streaming product. Prior to her stop at CNN.com, Smith spent three years as a senior producer with the Weather Channel and nearly 19 years as a supervising producer with CNN/*Sports Illustrated*, a 24-hour sports news network and a producer with CNN Sports. Smith earned a bachelor's degree from Central Michigan University.

Joan Van Tassel (Ph.D., University of Southern California) is an Associate Professor and the Chair of the Department of Communication Arts at National University. She is an educator, author, and journalist in new media, broadcast and print. Dr. Van Tassel has written five research-based books about key business issues in the information, telecommunication, and entertainment industries brought about by digital technologies and networks. Her most recent published book is *Digital Rights Management and Beyond*, published by Focal Press in 2006. She is currently working on *Media Management: Making, Marketing, & Moving Digital Content*, also for Focal Press.

At National University, Dr. Van Tassel teaches courses in interactive media, content distribution processes and technologies, media management, and persuasion in traditional and interactive media. Her research interests include the effects of interactive media on traditional media, pedagogical methods for adult online learners, and representation in media. Before coming to National University, she taught at Pepperdine University and the UCLA Extension School. Her doctoral degree is in Communication Theory and Research, from the Annenberg School for Communications, Los Angeles (1988). She has won numerous awards, including two coveted prizes for her books and an Emmy nomination for her work in television. Most recently, Dr. Van Tassel's 2009 article on the Obama campaign's use of digital technologies and new media platforms won first place from the San Diego Press Club for a magazine article on politics.

James R. Walker (Ph.D., University of Iowa) is Professor and Chair in the Department of Communication at Saint Xavier University in Chicago. He has co-authored four books on the U.S. television industry: *Centerfield Shot: A History of Baseball on Television* (2008), *The Broadcast Television Indus-*

try (1998), *Television and the Remote Control: Grazing on a Vast Wasteland* (1996), *The Remote Control in the New Age of Television* (1993). Dr. Walker has published more than 30 articles on various topics in mass communication. His scholarly interests include electronic media programming, new video technologies, and sports media.

Maria Williams-Hawkins (Ph.D., The Ohio State University) is an Associate Professor in the Department of Telecommunications at Ball State University. She worked in both commercial radio and television in Memphis, Tennessee, before going into education. She worked as News Director and Executive Producer at KMOS-TV in Warrensburg, Missouri, and went on to teach at Sam Houston State University where she taught production, special topics and news, with a focus on coverage of death row inmates. Dr. Williams-Hawkins also operated an independent production company. At Ball State University, her teaching includes news, production and international media management, focusing on China. She has published on international media economics and an analysis of Lifetime Television and presented papers on media representations of age, ethnicity, gender and inclusivity. She has produced and co-produced over 100 public affairs television programs on ethnicity, gender and international topics. Dr. Williams-Hawkins is an ordained minister and has served on both religious and international editorial boards. In 2009 she received the university's first Diversity Advocate Award.

Lily Zeng (Ph.D., Southern Illinois University Carbondale) is an Assistant Professor in the College of Communications at Arkansas State University, where she teaches mass communication and interactive media technology and conducts research in the use of interactive media content and devices. Dr. Zeng's research has been presented at national and international conferences and appeared in publications including *Journal of International Business Disciplines, Business Research Yearbook,* and *Financial Services Review.*

CPSIA information can be obtained at www.ICGtesting.com
Printed in the USA
BVOW020721251111

276782BV00003B/3/P